The
Kitchen Garden
Yearbook

The Kitchen Garden Yearbook

A Month-by-Month Guide to Growing Your Own Vegetables

Daphne Ledward

ROBSON BOOKS

Illustrations by Freda Ward

The photographs on pages 101, 116, 118, 121 courtesy of Thomson and Morgan
Book design by Harold King

First published in Great Britain in 1998 by Robson Books Ltd, Bolsover House, 5–6
Clipstone Street, London W1P 8LE

British Library Cataloguing in Publication Data
A catalogue record for this title is available from the British Library

ISBN 1 86105 167 0

Typeset by Columns Design Ltd., Reading

Printed and bound in Great Britain by
Butler & Tanner Ltd, Frome and London

INTRODUCTION

This is a record of the cropping of a piece of land 24ft deep by 45ft wide over a twelve-month period. It describes in detail both the successes and not-so-successful results, but above all it proves that it is possible for a household of reasonable size to become self-sufficient in essential vegetables from a plot of this size in a minimum time from the start of the project.

With one minor exception, no chemicals were used for the purposes of weed, pest and disease control, all such controls being effected by mechanical and barrier means. Inorganic fertilizers were used where necessary, but soil texture and fertility were improved by the addition of a large amount of home-made garden compost during the winter before planting.

In addition to the main piece of ground, an otherwise unused headland was planted with courgettes, as much for efficient ground cover as for the crop produced. Young plants of certain vegetables were raised in an unheated domestic greenhouse where necessary, but if this had not been available, it would have been possible to have produced them using a light windowsill in the house and a coldframe or cloche structure.

Because of changing prices of seeds and other relative commodities, no comparisons have been made between home-grown and shop-bought produce. However, a similar project undertaken three years earlier, where the cost of seeds, fertilizers and accessories was strictly recorded and compared with the annual greengrocery bill, showed that within twelve months the gardener should have broken even. From then on, taking into account the cost of accessories such as canes and crop protection covers over a seven-year period, a considerable saving would be made. This benefit is in addition to those of better flavour and increased nutritional value offered by home-grown, freshly harvested vegetables.

One way of maximizing the economic viability of such a project is to ensure there is a minimum amount of waste. I have therefore included recipes, cooking and serving suggestions for not only the mature crops but also for thinnings and parts of the plant normally discarded during preparation. Using as much of the vegetable as possible is particularly important during the early stages of the venture, where time is of the essence as far as self-sufficiency is concerned.

THE PLOT

The piece of ground used for the project is situated in south Lincolnshire. The climate is even, owing to the influence of the Wash, with few severe peaks and troughs of temperature on a regular basis, although the area is subject to cold easterly winds early and late in the year. Sowing times are therefore about average for the United Kingdom as a whole, though those living in very mild, sheltered localities and particularly cold areas may have to adjust their timings to suit their particular situations.

The soil is fen silt which has a tendency to become over-moist and consolidated during prolonged wet periods. Fine seed beds are liable to 'panning' (developing a hard crust on the surface) after watering, which can cause problems with germination, especially during cold periods. Its light brown colour may cause a delay in spring warming, so early sowing requires the soil to be prewarmed under polythene or other protective sheeting.

Initially the plot was very open and exposed to the elements, but five years ago shelter belts of British native trees and shrubs and fast-growing willows were planted. These now afford considerable protection, the temperature inside the belts being consistently several degrees higher than in the fields beyond. Most shrubs and willows are now coppiced on a regular basis to provide a dense undercover to the hardwoods, which at the time of the project were around 25ft (22.5m) high and provided a highly efficient windbreak for the whole of the kitchen garden.

Silt, like clay, is potentially fertile as the texture retains nutrients efficiently. The addition of large quantities of organic matter can help the dense nature of the soil, and this particular piece of ground had been well-manured with garden compost and horse manure for the previous six years during its time as a vegetable trial ground, which considerably reduced the initial problems I encountered. All perennial weeds had long since been eliminated through careful cultivation, though the nature of the soil being what it is, annual weeds grow with alacrity virtually year-round if allowed to do so.

THE SHOPPING LIST

A summary of varieties grown

*Seeds and sets supplied by Thompson and Morgan, Poplar Lane,
Ipswich, Suffolk IP8 3BU*
Potato varieties available from good garden centres

Rhubarb Chard

Early Broad **Bean** *Express*

Broad Bean *Imperial Green Longpod*

Dwarf Bean Mixture *Aramis, Purple Teepee, The Prince, Radar*

Climbing Beans Mixed *Romano, Desiree, Lady Di, Red Rum*

Beetroot *Mondella*

Broccoli Mixed *F1 Green Comet, Romanesco, F1 Caravel,* Extra Early
Purple Sprouting *Rudolph*

Brussels Sprouts *F1 Icarus*

Cabbage Mixed *F1 Perfect Ball, Bingo, Tundra, Tarvoy*

Carrot *F1 Mokum, F1 Bertan*

Cauliflower *F1 Elby, Walcheren Winter Pilgrim*

Courgettes **Salad Collection** *De Nice à Fruit Rond, F1 Gold Rush, F1
Greyzini, F1 Sardane*

Leek *Winora*

Lettuce *Fresh Salad Mixed*

Onion **Sets** *Jet Set*

Shallots *Atlantic, Pikant*

Spring **Onion** *Santa Clause*

Parsley *Afro*

Parsnip *F1 Gladiator*

Garden **Peas Mixed** *Daybreak, Hurst Green Shaft, Oregon Sugar Pod,
Sugar Snap*

First early potato *Rocket*

Second early/early maincrop potato *Nadine*

Radish *Rainbow Salad Mixed*

Sweetcorn *F1 Honey Bantam Bicolour*

Tomato *F1 Sungold*

Turnip *F1 Tokyo Cross*

ACCESSORIES

Enviromesh crop protection sheets – a fine mesh netting with a life expectancy of around seven years, designed to defend crops against frost, wind, hail, heavy rain, insect pests, birds, rabbits, squirrels, cats and dogs, and encourage earlier maturing.

Horticultural fleece – a lightweight, spun bonded material used to protect seedlings and plants from bad weather, bring on early crops, increase humidity and provide a barrier against pests and some airborne diseases.

Permealay ground cover fabric – a black, permeable fabric which can be used to warm soil, control weeds and protect against damage by many root pests, while allowing rain and other moisture to penetrate freely.

Canes or poles for climbing beans and tomatoes. Brushwood or netting to support peas. Soft string for tying tomatoes to canes. Growmore fertilizer.

Canes and netting supporting beans and peas

SELECTING THE VARIETIES

In choosing which vegetables to grow, I decided to opt for ones I use in large quantities rather than those, like Florence fennel and celeriac for instance, I may only buy once or twice a year as a special treat. I selected varieties for quality rather than quantity, as shop-bought ones tend to be grown primarily for yield, flavour being a lesser consideration. This has produced a mixture of reliable established names and newer types, often F1 hybrids, which can offer distinct advantages over older forms.

F1 hybrids are the first generation hybrids arising from the deliberate crossing of two carefully selected parents. Because this kind of hybridization is labour-intensive, F1 hybrid seed tends to be more expensive, but high quality and uniformity are reliable, which generally makes the extra cost worthwhile.

It can be expensive to grow a wide range of a particular vegetable in a comparatively small plot, as you generally end up with a large number of half-empty packets which seldom get used up the following year. On the other hand, raising just one or two forms of one kind of vegetable can be monotonous both for the gardener and, ultimately, the cook.

One way round this is to avail yourself of some of the vegetable seed collections offered by many companies. These either consist of several smaller, named packets of different varieties, or if more convenient, a mixture of various types within the same packet. This ensures a wide selection while providing seed in sensibly small quantities.

Rhubarb chard

This is an attractive, easily grown and more versatile alternative to spinach. The red mid-rib is removed and cooked like asparagus and the rest of the leaf can be used in all recipes calling for spinach.

Broad beans

Express   A prolific, well-flavoured, quick-maturing broad bean with white seeds. It freezes particularly well.
Imperial Green Longpod   An old favourite which is difficult to beat. The exceptionally long pods contain up to nine green seeds which are equally good used fresh or frozen.

French beans

These fill the summer gap before runner beans arrive. They freeze well and produce crops over a long period if the pods are regularly picked. A wide range of flat-podded, round-podded and coloured-podded forms is available, and buying a mixture such as that offered by Thompson and Morgan gives an opportunity to sample four quite different varieties within a small space.
Aramis   This is an attractive 'filet' bean with purple-flecked pods. The flavour is excellent, cropping is heavy, and the plants are resistant to several diseases.
Purple Teepee   The purple pods are easy to pick as they are very noticeable and are borne at the top of the quick-maturing plants.
The Prince   Still popular, this established flat-podded form has a superb flavour and freezes well.
Radar   A multi-podded gourmet variety with a succulent flavour. The pods should be picked when they are about 4in (10cm) long and cooked whole.

Climbing beans

These are an excellent way of utilizing a plot economically as a lot of beans can be produced in the airspace above a single, narrow row. As with French beans, a climbing bean collection enables you to try smaller quantities of several distinct varieties.

Purple Teepee –
*its striking
purple pods are
easy to pick*

Romano   This is a climbing French bean which looks and tastes totally different from both dwarf French beans and runner beans. The pods are long, tender and fleshy and the yield is high.

Desiree   A stringless bean with white flowers unattractive to birds and meaty pods with few seeds, it is a variety which crops well even in dry conditions.

Lady Di   The long, tender, well-coloured pods and the long season make this a popular choice even if only one runner bean variety is to be grown.

Red Rum   A new, hybrid runner bean with large bunches of pods – which are easy to spot amongst the sparse foliage – and good disease resistance.

Beetroot

Mondella Modern beetroot varieties require less thinning as, unlike older types, each seed only produces one plant. *Mondella* is an early maturing form which will remain in the ground over a long period without running to seed (bolting). The medium, even-sized, roots are a lovely deep red colour, even after cooking.

Broccoli

This has become a very fashionable vegetable in recent years, with the result that a bewildering number of varieties is now on offer to the amateur gardener. As most are 'cut and come again' forms, it is only necessary to grow a few plants, but to be able to pick broccoli over a long period, several varieties are necessary, and this is where collections of seed become useful again. Broccoli freezes exceptionally well, and so surpluses can be retained for future use.

F1 Green Comet This is an early calabrese type such as is usually seen in the supermarket. Large, deep green heads are produced in summer and early autumn, and when these are removed, there will be a secondary crop of small spears.

Romanesco The first of a new generation of calabrese with pointed heads, *Romanesco* took both cooks and gardeners by storm when it was first introduced. It is still among the best flavoured and most reliable of these types, following on in autumn and early winter from the summer varieties.

F1 Caravel This variety is slightly later than *Green Comet* and earlier than *Romanesco*, ensuring a continuity of supply of broccoli. It has very large, green heads, an abundance of smaller side shoots, and plants can be grown close together, so the resultant heavy yield makes this an ideal variety for a small garden.

Extra early sprouting broccoli *Rudolph*.   Sprouting broccoli is winter hardy. Early varieties usually start to produce spears from the beginning of January, and continue cropping until spring if regularly picked, when the ground can be prepared for the new season. *Rudolph* is well-flavoured, and the spears are large and of an intense red-purple colour.

Brussels sprouts

F1 Icarus   No winter menu would be complete without Brussels sprouts. The flavour of home-grown ones bears no resemblance to the tired, shop-bought specimens. Sprouts are available from September to June depending on variety, but as they are a fairly space-consuming crop, it is usually best to choose just one sort which will give you a crop over as long a period as possible. *Icarus* crops from October to February, which is probably as long as you would want to eat sprouts for anyway, and as this variety freezes well, you can always save a few pounds in case you get a craving for Brussels sprouts in summer!

Cabbage

I suppose school dinners have ruined cabbage for most of us over a certain age but it is well worth giving interesting varieties of home-grown cabbage a second chance, especially as there are so many exciting ways of serving (or, in the case of my husband, disguising) this useful vegetable. It is possible to produce a crop of cabbage at virtually any time of the year, but as this is a vegetable which should be picked and cooked at its peak of perfection, it is not necessary to grow vast numbers of plants and here, again, a seed collection of varieties comes in handy.

F1 Perfect Ball   Probably one of the best of the summer cabbages, well-shaped, uniform and a good flavour both cooked and raw in coleslaw. Although early to mature, it will stand for over three months without deteriorating.

F1 Bingo   A follow-on from *Perfect Ball*, *Bingo* produces large, tasty heads and will store in nets in a cool place for several weeks after cutting.

F1 Tundra   Excellent for winter salads and hot dishes, *Tundra* is extremely hardy and will stand throughout the worst of winters without spoiling.

F1 Tarvoy An outstanding savoy cropping from December to March, with dark-green, crinkly leaves and a wonderful flavour, especially when served with butter and a sprinkling of tarragon.

Carrots

It would be unthinkable to have a kitchen garden without carrots. Anyone who has only sampled frozen and shop-bought commercial carrots will have no idea of the true flavour of this essential vegetable, especially as the newer garden varieties have been bred with taste in mind.

F1 Mokum An early, conical, high yielding carrot which is suitable for eating cooked, raw in salads, and making into juice.

F1 Bertan This is a heavy cropping maincrop with a particularly good flavour and resistance to splitting. It will remain in the ground until the new year without detriment and freezes very successfully if necessary.

Cauliflower

Where would we be without the cauliflower? For hot dishes, as a side vegetable, in stir-fries and salads, its crisp nuttiness is unrivalled. The two varieties I eventually chose should provide plenty of fresh and frozen curds throughout the year, and modern breeding has ensured that many of the problems associated with growing cauliflowers in the past have now been eliminated.

F1 Elby This is a summer cauliflower with high tolerance to both drought and wet conditions and large, creamy heads of high quality.

Walcheren Winter Pilgrim One of the latest in the Walcheren breeding programme, combining winter hardiness with a reliable crop of compact, creamy-white curds.

Courgettes

Courgettes can waste a lot of space in a small garden, but are ideal for container and growing bag cultivation. Two or three plants are as many as you generally need, because although they can be frozen, the flavour and texture are nothing like those of the courgettes picked off the plant and cooked immediately. However, in this case, as their large leaves and dense habit make such good ground cover, I decided to use them to keep the weeds down in two areas of garden not included in the main vegetable patch which I did not want to use for anything else in particular.

Courgettes Salad Collection   I chose this selection of four very different varieties from Thompson and Morgan in order to grow a range of totally different-looking types. I use courgettes raw in salads as much as in hot dishes, and so this diversity of appearance is especially useful.

De Nice à Fruit Rond   A round courgette with a unique flavour, it is best picked at golf ball size.

F1 Gold Rush   This is a heavy cropper with a golden yellow skin. It looks great served on its own or mixed with green-skinned courgettes.

F1 Sardane   A dark, thin skin and huge cropping potential make this one of the best modern varieties.

F1 Greyzini   Originally known as *Blondy*, *Greyzini* is pale green in colour and is a reliable cropper as nearly every flower is female and produces a fruit. Larger fruits can be used as marrows, and mature ones will store in a cool place for several weeks.

Leek

Winora   As with Brussels sprouts, by choosing the right varieties and timing the sowings, there are few months in the year when leeks cannot be available. However, as it is a vegetable which lends itself admirably to hot winter dishes and cold winter salads, it is probably best to choose a variety which gives a long harvesting period during the colder months of the year from one sowing. *Winora* can be lifted from December to April, and produces thick, medium length shanks topped with handsome blue-green leaves. Bolting is rare, and the plants are rust tolerant.

Lettuce

Lettuce varieties are very much a personal taste – many people like the soft butterhead types and cannot stand hard, crisp, iceberg forms, and the looseleaf, pick-and-pick-again sorts are generally either enjoyed or loathed. For this reason, I tend to grow lettuces from a mixed packet as these are inexpensive, give a huge selection of forms and flavours, mature over a long period, eliminating much of the need for successional sowing, and provide much interest as the different varieties manifest themselves.

Onions and shallots

Over the years I have increasingly become convinced that the easiest and most successful way to grow onions is from heat-treated sets rather than seed. Although this appears a more expensive way of raising a crop, the end results are usually better. Until recently, shallots were available only as sets, but now seed is also available. However, I still believe that the best crop of shallots for the average amateur, like onions, is to be had from sets.

Onion Jet Set has largely superseded the excellent *Turbo*, which is still available from some seed companies. These onions, unlike many raised from sets, are round, a good shape and size, and mature earlier than most other varieties.

Shallots *Atlantic and Pikant*   Shallots have a milder flavour than onions and their size makes them good for using whole in dishes which call for it. *Atlantic* is a golden-brown skinned variety with white flesh, while *Pikant* is red, with a red tinge throughout the bulb.

Spring onion *Santa Clause*   Spring onions are invariably raised from seed and, because they are not required to produce bulbs, are easy to grow. *Santa Clause* is quick to mature so can be used early in the season, but will thicken up over the summer and can eventually be used as early leeks. The outer skin towards the base has a red tint which becomes more pronounced later in the year and can be encouraged by earthing-up.

Parsley

Afro   I use such a lot of it that parsley is the only herb I grow in quantities large enough for me to classify it as a vegetable. *Afro* is a prolific, strong-growing variety with large, bright-green, curly leaves which can be picked throughout winter. It also freezes well for emergency use in cooking – no chopping is necessary as the leaves can be crumbled straight into the recipe!

Parsnip

F1 Gladiator   This is a hybrid parsnip which produces a reliable, well-flavoured crop with good disease resistance. Germination can be a problem and success with parsnips generally depends on sowing later than usually recommended, when the soil has warmed up.

Peas

There is nothing more delicious than fresh peas taken straight from the garden and lightly cooked, so the most important thing about growing them is to make sure you have crops for as long as possible, using a combination of early and maincrop varieties. The increasing popularity of mangetout and snap peas where the whole pod is eaten, not just the peas inside, makes the inclusion of one or two of these varieties worth considering. A seed company's garden pea collection, such as the one in this project, will usually contain a balanced mix of varieties to satisfy most tastes over a long period.

Daybreak   Probably the earliest pea which should start cropping in late June or early July from an early spring sowing. The pods are large and contain more peas than other early types and the flavour is very good.

Hurst Green Shaft   Over all the years I have been growing peas I have not come across a variety to better this one. The pods are long and abundantly produced in pairs, and contain an average of ten sublimely flavoured peas per pod. There is a long cropping season usefully following straight on from *Daybreak*.

Oregon Sugar Pod   This is a mangetout variety with good disease resistance, producing a heavy crop of flat pods which can be frozen successfully if required.

Sugar Snap   A popular snap pea with thick, juicy pod walls, remaining sweet and tender even when very mature.

Radish

Rainbow Salad Mixed   To my mind, radishes are vital for the best mixed salads. For good appearance a variety of sizes, colours and shapes should be grown. This mixture contains a good range of red, pink and white forms both cylindrical and round, which mature at different times to ensure a long cropping period from one sowing, thus preventing the need of sowing for succession which is necessary if just one variety is grown.

Sweetcorn

F1 Honey Bantam Bicolour   The last few years have seen sweetcorn move from a little-used gourmet vegetable to an essential part of cookery. Modern breeding has produced early ripening varieties, which have taken much of the hit-or-miss business out of growing sweetcorn at home, especially for northern gardeners and in bad summers, and the newer, sugar-enhanced varieties produce kernels which are so sweet and tender they can be eaten raw. *Honey Bantam* is very quick to mature and its 7in-long cream and gold bicoloured cobs look delightful when served running in butter.

Tomato

F1 Sungold   Recommended for both greenhouse and outdoor cultivation, this is one of the increasingly popular cherry tomato types with arguably the best flavour of any available to date, being both super-sweet and tangy. Its huge yields of cascading trusses of striking orange fruits bring as much pleasure to the eye as to the palate, and cropping should continue well into the autumn until the first really hard frost.

Turnip

F1 Tokyo Cross   Turnips are not everyone's cup of tea, but they are undeniably useful in soups, stews and casseroles, and they do liven up carrots if the two vegetables are mashed together with plenty of butter and a sprinkling of dill seed. *Tokyo Cross* is a quick maturing variety providing attractive, uniform roots with a tangy, nutty flavour which are best harvested small and cooked whole, but can be

Potatoes in a trench

allowed to grow bigger for dicing. It freezes well and keeps its flavour and texture over a long period.

Potatoes

Potatoes are a land-consuming crop, and it can be argued that the space they occupy and the price of the seed does not justify their inclusion in a small vegetable garden. It is good to be able to choose your favourite varieties, however, and a few first earlies and second early/early maincrops will give you the taste of really fresh new potatoes before the market is flooded with the popular, albeit excellent, commercial maincrop varieties. For this project I used 12 self-saved seed potatoes of each of the two early varieties I have found to give the best all-round performance for amateur cultivation, mainly because space would not allow me to grow more. I would not ordinarily recommend growing self-saved seeds, as disease problems are more likely to occur than with specially raised certified stock, but it is virtually impossible to buy such small quantities of seed potatoes and providing new stock is obtained regularly and the crop is rotated, on occasions like this the risk is worth it.

First early potato *Rocket*   A newer variety and one of the earliest potatoes around, with good eelworm resistance. It is a heavy cropper, producing good-looking, uniform, white tubers which cook well in a variety of ways.

Second early/early maincrop potato *Nadine*   Another new variety which deserves to be more widely grown. It is a red-skinned variety with a waxy texture which boils, bakes and roasts well. In addition it is disease-resistant and also shows some resistance to slug damage.

19

FIRST YEAR VEGETABLE PLOT AT-A-GLANCE CALENDAR

Month	Early	Middle	Late
January	Rough-dig parts of plot not already dug. Remove perennial weeds and bury annual ones. Add compost or well-rotted manure. Order seeds and sets.		Put seed potatoes in a cool, light place to sprout (chit).
February	Cover soil with polythene or Permealay during a mild spell to pre-warm it before sowing. Start off shallots in seed trays of compost under cold glass.		Sow summer cauliflowers in seed trays under glass.
March	Sow Brussels sprouts under glass.	Start off onion sets in seed trays under glass. Spread balanced fertilizer on rough-dug soil, level soil and recover. Prick out summer cauliflower seedlings into small pots under glass. Sow parsley and leeks in seed trays under glass. Plant shallots in pre-warmed soil outdoors.	Plant first early potatoes. Sow cabbage plants in seed trays under glass. Sow broad beans and peas into pre-warmed soil. Prick out Brussels sprouts seedlings into small pots under glass.
April	Prick out leeks into boxes under cold glass. Sow rhubarb chard, carrots, spring onions, lettuce, radishes. Plant second early potatoes. Sow courgettes in small pots under glass.	Prick out cabbage seedlings into small pots under glass. Provide support for peas. Sow tomatoes in seed trays under glass.	Sow parsnips. Plant out summer cauliflowers. Sow broccoli in seed trays and sweetcorn and climbing beans in individual pots under glass. Plant out parsley seedlings.
May	Prick out broccoli and tomato seedlings into small pots under glass.	Thin lettuce seedlings when large enough to use in salads. Start pulling radishes. Sow F1 turnips. Provide supports for broad beans. Sow French beans in situ. Plant out cabbages and Brussels sprouts.	Plant out climbing beans, sweetcorn, courgettes.

Month	Early	Middle	Late
June	Plant out broccoli. Plant tomatoes outdoors. Plant out leeks. Use turnip thinnings as 'greens'.	Cut first mature lettuce heads. Remove broad bean tops once crop has set to deter blackfly. Start picking early peas. Sow winter cauliflowers under glass.	Start harvesting F1 turnips. Use turnip tops as a green vegetable. Begin picking rhubarb chard and 'snap' peas.
July	Use early carrot thinnings as 'finger' carrots. Start harvesting early broad beans.	Start lifting early carrots and early potatoes. Begin picking French beans, maincrop broad beans and peas and courgettes.	Thin beetroot. Use thinnings as 'baby beet'. Cook tops like spinach. Commence cutting early calabrese and summer cabbage. Plant out winter hardy cauliflowers. Start harvesting summer cauliflowers and climbing beans.
August	Start picking tomatoes and climbing beans. Thin maincrop carrots and use thinnings as 'finger' carrots.	Start lifting second early potatoes. Dig areas left empty by harvested crops. Harvest onions and shallots.	Start digging maincrop carrots.
September	Start harvesting early autumn calabrese. Harvest sweetcorn.		
October	Plant winter hardy onion sets.	Start cutting autumn cabbage.	Begin picking Brussels sprouts.
November	Begin harvesting late calabrese.	Start lifting parsnips.	
December	Start lifting leeks and cutting winter cabbage.		Begin cutting savoys.

Harvest sprouting broccoli from January, and winter hardy cauliflowers from late April.

THE DIARY

December

Saturday 21 and Sunday 22

Two lovely sunny days, not a cloud in the sky from dawn until dusk. I was months behind on the vegetable garden owing to pressure of work, and much of the remains of the summer vegetables were hanging around, so this seemed a good time to clear them up and get the plot ready for the winter. There was a certain amount of urgency as both nights there was a slight frost. Frost is needed to work on the soil over winter, but it is not a good idea to bury frosty soil, which can remain cold in spring much longer than that near the surface.

After removing old tomato plants, defunct leaf beet and old bean haulms, I lifted the rest of the carrots and beetroot – and then the bulk of the land was ready for digging. I had four bins of well-rotted garden compost and this was spread about 6in (15cm) thick in the bottom of 15in deep trenches throughout the vacant parts. This has been the practice for the last five winters, using either compost or farmyard manure or both, and has resulted in a deep, humus-rich topsoil. The object of burying the organic material rather than incorporating it into the top layer of soil is twofold: it encourages deep digging and ultimately a greater depth of topsoil, and it enables me to add bulky manures every year to every part of the plot. This, in

A humus-rich topsoil is the result of spreading compost in deep trenches (right). *I keep several bins of compost* (far right)

turn, has opened up the close texture of the soil enormously, while avoiding the problem of forking of root crops which can occur if they come in contact with fresh manure or compost.

I chopped up the old plant remains and used them as the basis of my next supply of compost.

There is a small area still containing brassicas and leeks which will have to be cleared and dug after Christmas when the crops have been harvested. Otherwise I am now well on course for a new season hopefully laden with tasty vegetable goodies and hooray, the shortest day has come and gone!

From 23 December until 12 January there was continuous frost with periods of lying snow. This was ideal for the land already cleared and dug, but prevented any further work on the remaining parts.

Tuesday 14

A welcome mild day, when everyone was declaring it felt good to be alive, and a post-Christmas need for exercise urged me to finish clearing the vegetable patch. I lifted the last of the leek and parsnip crop, picked the remains of the Brussels sprouts, and cut the final cabbage heads. I will store the latter in the garage, where they will be more handy for me than three miles down the road – the somewhat inconvenient distance between home and the kitchen garden. Digging and manuring the vacated area got the blood coursing nicely, and the sight of the well-worked land waiting expectantly for the spring sowings was most satisfying.

Wednesday 15

It is as well to keep a check on the acidity or alkalinity of the soil when a lot of organic material is added to it on a regular basis. A soil which is too acid grows few vegetables really well, but on the other hand indescriminate liming can make the soil so alkaline that only members of the cabbage family will thrive. Despite a deterioration in the weather and a whole day of freezing fog which did not encourage me to leave the fireside, I decided to do a pH check on the soil. This showed that for the first time conditions were becoming acid, so I spread lime over the surface at 8oz/sq yd (240g/m^2) and left it to the elements. Manure and lime should not be allowed to come in contact with each other, as valuable nitrates will be released from the manure and lost into the air as ammonia. However, as my garden compost is well buried, and any old organic matter is long since spent, there should be no problem here.

Thursday 16

A bitterly cold, foggy and frosty day provided a good opportunity to stay indoors and compile a seed order, always an enjoyable job as there is much pleasure to be had in anticipation. Many of the varieties I have ordered are the same as last year, which yielded a bumper harvest. I made one or two substitutions where I felt I could do better or where I thought I could utilize the limited space more profitably.

Sunday 19

Welcome rain fell solidly for 24 hours, settling soil lifted by the frost and clearing the filthy roads and paths of salt and mud. It was an ideal day to work out a three-year crop rotation and draw a scale plan of the first year's crops. Although a scale drawing is not essential, it saves a lot of bother at sowing time, and it does make sure in advance that you can fit the vegetables you want to grow into the space available.

Crop rotation is necessary to avoid a build-up of pests, diseases and other disorders. Rotation is generally done over a three-year period and involves dividing the plot into three areas, each containing crops of a similar nature. Following certain types of crop by others of a totally different constitution makes it easier to add appropriate amounts of manure, fertilizer and lime to each of the three sections when necessary.

Seeing the vegetable garden laid out on paper has a positive psychological effect, and when I had finished I could almost taste the new season's crops already!

Wednesday 22

Hooray, the seed order has arrived – or most of it, anyway. On checking, I found that the sweetcorn, onion sets and shallots are to follow but there is no panic, as I can't start planting outdoors for another two months. The arrival of the seeds was just what I needed to lift my flagging spirits on a cold, wet day when daylight seemed conspicuous by its absence.

Sunday 26

Time for starting my seed potatoes into growth. I saved 12 each of first early *Rocket* and second early/early maincrop *Nadine* when the crop was lifted last year. These are about the size of a hen's egg and contain a good number of eyes (dormant buds) at one end. I have placed them in a seed tray in a light position a little way back from the east-facing window of our utility room, which is usually unheated and maintains an even temperature of around 50°F (10°C). This should encourage the production of a number of strong, green shoots by planting time.

Seed potato with a good number of dormant buds

Thursday 30

The shallots and onion sets arrived in the post this morning. Anyone who has not bought shallots before would probably be surprised at how few there are in a bag – 8 *Atlantic* and 12 *Pikant*, including some small bulbs which had broken off from the main ones. However, from experience, this will probably provide as many shallots as required – their shape and mild flavour make them useful for pickling and certain dishes calling for a hint of onion or whole, small bulbs. For general use, the *Jet Set* onions should be enough to see us through to the early Japanese onion crop next July.

February

Saturday 1

A good way to spend a wet day was to make television programmes on fruit and vegetable growing for the complete novice. Much time was spent persuading the researcher that the pansy seeds she had brought in to demonstrate sowing vegetable seeds were, in fact, not vegetables. The make-up girl was also in a spot of bother, this time because she had just moved into a new house with a small vegetable plot and her young son had just sown his sweetcorn seeds in a pot on the kitchen windowsill. She was somewhat dismayed when I pointed out that, having sown them so early, by the time it was warm enough to plant them out in Manchester they would be several feet tall.

Sunday 2

Many people have trouble growing shallots as they can run to seed in certain seasons. The reason for this is that they have often been planted too late in the spring, resulting in stress during hot or dry spells before the bulbs have become sufficiently well established to cope. It is important, therefore, to plant them as early as possible, but this can be a problem in cold springs, for if the soil has not started to warm up, they will just sit there and sulk. There are two tricks you can use to get round this. One is to pre-warm the soil; the other is to start the shallots into growth before they are planted out in the garden.

We have had about a week of fine, relatively mild weather, so today I felt it was time to get the soil warmed up. Covering it with almost anything – polythene, fleece or other crop protection sheeting, even old carpets! – will reduce wind chill and keep much of the late winter frost out. I had a small quantity of new Permealay sheets in the shed, so I used these to cover some of the area, and blanketed the rest with Enviromesh netting (see Accessories, page 5). It will be interesting to see which covering works best. The black sheet will absorb heat but will also radiate it during colder periods, so it may lose some benefit at times. The white Enviromesh will reflect a lot of the sunlight and so not warm the soil as quickly, but it will not radiate any absorbed

Warming the soil under black and white covers

Growing crops under Environmesh netting

heat during cold spells and it should break the effects of chilling winds and frost. In theory, therefore, the long-term effect should be about the same. Watch this space, as they say.

My unheated greenhouse now also warms up nicely during the day, so I decided to lose no time and get the shallots moving. I half-buried them in just-damp, general purpose compost in two seed trays (one for each variety) and put them on a polythene-covered polystyrene sheet to keep them snug, placing them on the main bench in full light. Over the next few weeks they should begin to form roots, and in about six weeks they can be carefully removed from the trays and planted out in the (hopefully!) pre-warmed soil. I am fortunate in having the facilities of the greenhouse but, if not, I would have used the unheated utility room, where I have started off the seed potatoes.

Monday 3

The bushes in the shelter belts surrounding the vegetable plot and adjoining soft fruit garden and orchard have put on an incredible amount of growth in the two years since they were coppiced and it is time to cut them back hard again. It was a beautiful morning – too good to waste indoors despite a pile of writing to tackle – so I made a start. There is about a week's work in all; the shrubby willows and dogwood which were planted for winter bark effect are cut back almost to ground level; elders, hazels, sea buckthorn and various other species I reduce to around four foot high. I am gradually making the young trees which were planted as 'whips' or cuttings five years ago into standards, with a high crown to form a long-distance windbreak, whilst still allowing the undercover of coppiced shrubs to develop around them. Providing this work is done on a regular basis, a very efficient shelter for the vegetable and fruit gardens is created without too much loss of sun.

However, once I started the work I noticed how much I had allowed the shrubs to spread into the vegetable area over the last couple of years – about a yard at either side has been temporarily lost, but will shortly be regained. I must keep my eye on that area during the summer, to ensure that once the coppiced bushes start to regrow, they won't trespass again.

An enormous amount of rubbish is produced during this work. It is not feasible to burn it as there is insufficient room and the neighbours would object, so we have obtained a heavy-duty Makita electric shredder which has an insatiable appetite and makes short work of any smaller branches not worth saving for use on the log-burner in the cottage. This crushes woody tissue rather than chopping it, speeding up the rotting process. We use the material as a mulch underneath the trees and bushes from which it was originally removed, so nothing is wasted.

It is tremendously satisfying to stand back and admire the finished results, although it does take a little imagination to envisage how the area will look in twelve months, when the stumps have regrown and the brightness of the coloured bark on the young stems will be a delight to behold.

Tuesday 11

The coppicing is coming on nicely, and it is amazing how much early light is now reaching the vegetable plot. So far the weather has been very kind, mild and sunny, but we were rained off at 3pm this afternoon; however, there is only about another day's work left.

The shallots have already started to put roots down into the compost. They may be ready for planting out sooner than I had anticipated, but much will depend on the weather we get between now and then.

Shallots putting down roots into compost

Friday 14

The seed potatoes are beginning to sprout, not spectacularly, admittedly, but nevertheless they are looking promising, and there is another six weeks to go before the earlies can be planted, so we are well on target.

Tuesday 18

After nearly a week of rain and howling gales, the sun came out at lunchtime, and I seized the opportunity to finish coppicing and generally thinning the shelter belt on the east side. The wind was horrendously strong and very cold, but we managed to finish the shredding at 6pm by the light of a brilliant moon and the outside security lamp on the shed. It is a lovely feeling that the job is done, and very little work will now be required, other than keeping the trunks of the standard trees de-whiskered, for the next two years.

Wednesday 19

The potatoes are sprouting well, so I have moved them into the greenhouse where they will receive better light. Strong, dark green shoots – not the long, white, spagetti-like ones you find if you leave them too long in the dark – are essential if seed potatoes are to get off to a flying start in the ground. The temperature of the greenhouse varies between 34°F (1°C) at night and 68°F (20°C) during the day when the sun is out, so they should take no harm in there.

Friday 21

The weather is mild, and although still very windy outside, the greenhouse is snug enough for me to sow the first of the brassicas – summer cauliflower *F1 Elby*. F1 seed is expensive, and there are not many in a packet, but if every one germinates, there will still be more than twice as many as I shall need. My surplus goes to my friend Ruth, who has a big garden in the next village, but little time to raise vegetable plants from seed, so nothing is ever wasted.

Cauliflower seed is large enough to sow individually, so I spaced it out evenly in rows in a standard plastic seed tray of soilless, multi-

purpose compost and covered it with about half an inch of compost.
This should ensure strong, stocky plants for pricking out. From
experience, I find it is much better to sow F1 brassica seed in trays
under cold glass than in a nursery bed outdoors, where the small
quantity of seed tends to get lost. It also ensures an earlier start,
though each variety has its own maturing time, so one only gains a
week or so in the end.

March

Saturday 1

March is said often to come in like a lion, and this year is no exception. We could do with some gentle rain, but all we seem to get at present are gales and squally showers, interspersed with brilliant sunshine. The sunshine is useful in the greenhouse, as it is keeping temperatures at a reasonable level, and this morning I noticed that the *Elby* cauliflower seedlings are just beginning to poke through the compost in the seed tray. Inspired by this, I sowed the *F1 Icarus* sprouts in a similar fashion.

Sunday 2

Very windy again, but pleasantly sunny, so I ventured out to see how the soil is warming up under the covers. A soil thermometer is a useful and not particularly expensive piece of equipment and, although not essential to successful seed sowing, it does give a more accurate indication of when the time is right to start. I mentioned earlier that it would be interesting to know whether there is a difference if the soil is covered with white Enviromesh or black Permealay, so I checked the temperature beneath each and also in the open ground at the rear of the vegetable garden – the result: the thermometer read 48°F (9°C) in the soil beneath both covered areas, and only 40°F (5°C) where it had not been covered. The conclusion is that it does not really matter much what the soil is covered with, as long as it is given some protection from the elements in late winter and early spring. In fact, the covered soil has reached the temperature where it is possible to sow quite a few types of crop. I noticed that a few broad beans which had obviously dropped from the haulms when I was clearing up last summer had already germinated beneath the Enviromesh and had produced several large, healthy plants. Nevertheless, I am resisting the temptation to rush matters, as it is very early and we could still get severe frosts which could affect soil temperature, even under the covers.

Saturday 8

I visited a client's garden during the week before producing a set of plans for her. She is new to the area and wonders if she has done the right thing in moving to Lincolnshire from Suffolk, as she has never experienced so many unpleasantly strong winds in her life. I was able to reassure her that this was, in fact, quite unusual, and added that she should be thankful that they have generally been blowing from the west or south, as they often come from the east in this part of the country at this time of year. Anyway, the chill factor as far as the greenhouse is concerned is minimal, and the *Icarus* sprouts are now through the compost to join the cauliflowers, which are growing well and should soon be ready for transplanting into pots.

Sunday 9

It is time to add Growmore fertilizer to the soil so it can begin to release nutrients before planting starts. It was quite a job to remove the covers, as they filled like sails in the wind and we were in great danger of floating off in an easterly direction as we grappled to keep control.

The usual application before cropping is 4oz/sq yd ($125g/m^2$), which is worked into the top layer of the soil. This also helps to break down the clods so the ground will be ready for raking to a tilth immediately before sowing. I was delighted with the way the soil worked, and it looked really good before it was covered again – almost screaming out to be planted up.

It was interesting to see how many vegetables had inadvertently been left in the ground last autumn – in addition to the young bean plants, there were also parsnip roots which had started to shoot, and

even a huge onion that had survived the winter and was now sporting a fine head of green leaves. These had to be removed, of course, together with quite a covering of *Good King Henry* seedlings from some plants which live permanently on the headland. *Good King Henry* is a useful perennial vegetable and is handy for leaf salads and as a spinach substitute, but can become a nuisance as it seeds readily everywhere.

Monday 10

The *Afro* parsley I grew last year is starting to regrow. Although the plants look good and healthy at this time of year, and I am picking the leaves regularly, it is still advisable to grow a fresh supply every year. Parsley is a biennial, so it will flower the second year. The leaves are generally coarser and have a stronger and less pleasant flavour once flower spikes are produced, and most of the plants will die anyway at the end of the second season.

Many gardeners have trouble with growing parsley, especially with its germination, but I have to say that it has never been a problem with me. Tradition has it that parsley grows well where the woman wears the trousers in the home, so this may account for it! I always start parsley off in seed trays under glass as it is slow to germinate and it is much easier to keep an eye on it. Today seemed a good day for this job, as the young plants will then be about ready to replace the old ones when these are starting to flower. At the same time, I sowed a seed tray of *Winora* leeks, also in the greenhouse. This makes the seedlings much better to handle than if they are sown later in the spring in nursery beds outdoors, and will also give an earlier crop.

Tuesday 11

One of the hazards of being a 'media gardener' is that we have to spend large periods away from home just when work in the garden is particularly pressing. This year the problem is the Ideal Home Exhibition at Earl's Court in London, where I am working on the stand of a well-known gardening accessories manufacturer throughout the show, which lasts a month. Fortunately I do have days off when I can come home and keep things ticking over in the garden, but it is advisable to leave everything absolutely up-to-date before departing. This morning I noticed that white roots in the shallot trays were working their way to the surface, so I decided they must be planted out immediately, otherwise they would be in a terrible tangle by the weekend.

As it was, they separated quite easily. I planted them out, 1ft (30cm) apart, in holes along the row designated for them on the plan, which could be positioned quite easily by measuring off from either end of that section of the garden. They were firmed in gently to prevent damage to the mass of luscious roots, and then re-covered with Enviromesh. The soil is maintaining a temperature of about 50°F (10°C), so they should not suffer a check to growth.

Flushed with the success of the sprouting shallots, I decided to give my *Jet Set* onion sets similar preliminary treatment, and planted them 40 to a standard seed tray of compost (there were 80 sets in all, which should be enough for two rows if planted 6in (15cm) apart in the garden). By the end of the month, these, too, should have started to produce roots, which will get them off to a better start when planted out.

Saturday 15

Home from London for the weekend, I found that the cauliflower seedlings were ready for pricking out into small pots. I potted up 42 plants, 27 more than I shall probably need, but this will allow for accidents. The rest of the seedlings will go to Ruth when they are big enough. The Brussels sprouts are also nearly ready for pricking out, but would benefit from another few days' growth.

The leeks are pushing through the compost like grass, but it will be a little while yet before I can prick them out into boxes.

Sunday 16

The onion sets have already produced roots, so there is no point in waiting any longer to plant them out. This is an exceedingly early spring – about six weeks earlier than last year, which was a particularly late one, and one has to make the most of the weather. The problem many gardeners have with onion sets is that as soon as they are planted, the birds come along and pull them out again. Covering them with Enviromesh or similar crop protection sheeting will prevent this wanton vandalism.

I notice that the farmers round us are beginning to plant their early potatoes, so I think I shall shortly follow suit.

Monday 17

I am about to leave for the Ideal Home Exhibition again, so I potted up the *Icarus* Brussels sprouts – 35 in all. As usual, the surplus will be given away when they have re-established.

The parsley has appeared this morning – no problem with germination here! There is automatic watering in the greenhouse, but this is rather hit-and-miss when young plants are involved, so David, the husband of my friend Ruth, is keeping an eye on things while I am in London. I watered this morning, and if David checks the greenhouse on Wednesday, all should be well until I return on Friday evening.

Saturday 22

We have had another warm and pleasant week, and the *Rocket* first early potatoes have fat, dark green shoots on them, so into the ground they went, 1ft (30cm) apart, to fill half a row. The other half will take the *Nadine* second earlies later. I like to earth the row up a little at the start, so that the soil will protect early foliage from frost. The Enviromesh sheeting should also do this, so even if we get a late cold spell, there should be no problem in this respect.

Potatoes,
planted one foot
(30cm) apart
(above)
Sowing peas
Daybreak
(above right)
2in (5cm) apart

Sunday 23

It is still mild, so I decided the time had come to sow the peas and broad beans. The pea row was divided into four, each section containing equal numbers of *Daybreak*, *Hurst Green shaft*, *Sugar Snap* and *Oregon Sugar Pod*. The peas were sown at the base of a 3in (8cm) deep, flat-bottomed drill, spaced 2in (5cm) apart in three rows. The earliest, *Daybreak*, were treated with a fungicide to reduce disease – this would be a problem if one were trying to grow vegetables completely organically. Fortunately, I am not, but I must admit I am not keen on handling the dressed seed, which had a vivid red coating, and made sure I washed my hands thoroughly afterwards. There were a few seeds of each variety left – these will come in handy if there are any gaps in the row.

The broad bean row was divided in half, the first accommodating the very early *Express*, the second, furthest away from the front path, containing the old favourite *Imperial Green Longpod*. These were sown in a double row, 8in (20cm) apart, 2in (5cm) deep, with about 8in (20cm) between each bean. This used up all the beans in both packets. By the time I had finished, it was late, and the temptation was to leave the Enviromesh off until the morning, when I could see better. However, from experience, this would have been a very unwise move, as the pheasants would have descended immediately and dug them all out again. Two years ago I had to re-sow the entire

Honey Bantam – one of the newer generation of sugar-enhanced sweetcorn

pea and broad bean crop because I neglected to cover them up the same day they were sown.

Both the shallots and the onion sets are now sporting a good, healthy crop of leaves, and promise a heavy yield. Typically, the developing roots of the onion sets have forced the bulbs out of the soil, but the crop is covered, so I have left them alone for the time being, as no harm will come to them.

Monday 24

March has been a very dry month, but today we had a wet morning, which was excellent news. I could not work outdoors, so I took the opportunity to sow the cabbages in seed trays in the greenhouse. The rain will help the crops already planted outdoors, as the surface of the soil was beginning to look rather dry. I have resisted the temptation to water, however; at this time of year, Nature has a habit of correcting her little failings unaided – as she has today – and artificial irrigation is not a patch on natural rainfall.

Monday 31

This really has been a fantastic month weather-wise, and it is indeed going out like the lamb we had hoped for. I took a gamble this afternoon and planted my *Nadine* second early potatoes, about a week earlier than I would normally do, but the soil is warm enough, and I have to make the most of every spare minute I am at home. There is still another week of the exhibition to go. I shall celebrate its end by planting the bulk of my vegetable seeds, so I am praying that the weather holds.

All the cabbages are through the compost – at this time of year under unheated glass, now light levels and temperatures are much higher, the average germination time for brassicas is 5–7 days. Many of the early broad beans have also germinated, but as yet there is no sign of the peas.

Minimum soil temperatures for vegetable seed germination					
Vegetable	°F	°C	Vegetable	°F	°C
Broad bean	50	10	Marrow/courgette	65	18
French bean	55	13	Onion (seed)	50	10
Runner bean	55	13	Onion (sets)	40	5
Beetroot	45	7	Parsley	55	13
Broccoli	50	10	Parsnip	45	7
Brussels sprouts	50	10	Pea	38	4
Cabbage	50	10	Radish	38	4
Calabrese	55	13	Spinach/leaf beet/	45	7
Carrot	45	7	chard		
Cauliflower	55	13	Sweetcorn	65	18
Leek	45	7	Tomato	65	18
Lettuce	40	5	Turnip/swede	45	7

April

Tuesday 1

The seed tray full of leeks was bursting at the seams, and it was high time they were pricked out into boxes. In the shed I found two large, shallow, slatted wooden boxes used during the winter for storing apples. Lined out with black polythene bin liners, perforated in places to provide drainage, and filled with compost, these made ideal secondary containers for leek seedlings. There were around 150 in all, and at 75 per box there was plenty of space for them to develop. They will stay in the greenhouse for another month, before being hardened off outside, prior to planting in the vegetable patch. There will be enough plants, if spaced 4in (10cm) apart, for two rows. This is closer than often recommended, and does result in slightly thinner leeks but, wearing my chef's hat, I do prefer them like this to the fat, white, mushy ones often found in supermarkets, since both their flavour and appearance are vastly superior.

Winora *leek seedlings*

Wednesday 2

The weather is more like mid-May than early April. Some of the peas have now put in an appearance, and the onion and shallot tops are so well developed that I had to slacken the Enviromesh as they were getting bent. Because of the dry weather there are few weeds, and these are quickly knocked off with the hoe – it is better to deal with them as they emerge than leave them for a later blitz.

Thursday 3

Before leaving for a final stint at the Ideal Home Exhibition, I decided to sow the courgettes. The recommendation is to sow two to a pot and thin out the weaker seedling, but I often find there *is* no weaker one, and it is a terrible waste to discard a perfectly healthy young plant. I prefer to sow one seed in each pot, as I have plenty of room on the headland at the back of the vegetable patch to accommodate all of them, even if every one germinates. Of course, what to do with the resultant glut of courgettes and marrows is another matter, but I will worry about that later!

After London, I travel to Manchester for some Granada *Goodlife Garden Club* recordings, then home for about three precious weeks before taking to the road again.

Wednesday 9

The best laid plans of mice and men, etc – I had earmarked today to do my main outdoor sowings, instead of which I spent most of the day indoors feeling sorry for myself. Towards the end of the Exhibition, both John and I developed a flu-like affliction which left him with pleurisy and me with every 'itis' from the sinuses down. Despite the warm, sunny weather, I was in no state for hard labour, but did, however, summon up the strength to prick out the cabbage seedlings. Our various friends will do very well here. I potted up the strongest two dozen each of *Bingo*, *Perfect Ball*, *Tarvoy* and *Tundra*, and as I shall only require five of each there will be a fair-sized surplus.

Saturday 12

Our friends from Jersey are staying at a hotel about 25 miles away. This ancient coaching inn has a local reputation for comfort and good food and we were delighted to join them for dinner. What a disappointment! The food, so lavishly described on the menu, was mediocre in the extreme, the vegetables inedible – bland, wet mashed potato, filet beans so hard it was virtually impossible to get one's teeth into them, and a boring mixture of imported courgettes and tinned tomatoes described as 'ratatouille'. Why is it that British hotels try to impress with foreign vegetables cooked badly when, on their doorsteps, they have suberb, locally grown crops in season? It is ironic that ten miles along the road, across the county border in Lincolnshire, they could have bought cauliflowers, a variety of cabbage, spring greens, carrots and leeks. All these require is a chef with a knowledge of how to cook vegetables to provide both nourishing and delicious accompaniments. To add insult to injury, when Jersey Jenny (as she is known in our household to distinguish her from many other Jenny-friends) pointed out to the waiter that the vegetables were too hard to bite, he launched into an explanation of *al dente*, and inferred we were peasants who liked our veg boiled to a pulp. That was an extremely bad move – you do not try to tell Jerseymen (and women) how to cook, as the standard of cuisine there is flawless, regardless of size or type of eating establishment – which probably explains to some degree why John and I chose to marry there! As you can imagine, the air turned blue – especially when the bill came to nearly £200 for the four of us.

Tuesday 15

The flu-bug is winning, and today, a lovely spring day, turned out to be one when all I could manage was a visit to the doctor for antibiotics and bed rest – very frustrating as I really need to get sowing outside. To make me feel better, John sowed a tray of *Sungold* tomatoes in the greenhouse, and informed me that the first of the courgettes has made an appearance.

Thursday 17

I could leave it no longer. I had to get onto the Patch, despite still feeling like death and a bitterly cold wind. I shall be glad when the coppiced underplanting in the shelter belt has grown up a bit – the willows are already shooting nicely so it should not be long.

The first job was to remove the Enviromesh sheets, which revealed a crop of weeds, although not as bad as they might have been. It has not rained for nearly three weeks, and then there was not enough to do good. As the whole plot had already been dug over in the winter, it was sufficient to hoe off the weeds, which were all annuals like may-weed, groundsel and chickweed, and therefore would not regrow. The onions, shallots, broad beans and peas had to be hand-weeded, but I quite enjoy this job, and the crops looked so pleasing after-wards, with the exception of the *Hurst Green Shaft* peas which were virtually non-existent except for about four straggly plants. I had noticed some time ago that they did not seem to be making much of an appearance and assumed that, as they were early maincrop, they were probably merely late in germinating. It was now clear some-thing had gone wrong. Even scratching around in the soil failed to reveal any seed, germinated or not, so I can only assume it has been eaten, but by what? Certainly not birds, as the whole crop was covered. It could have been mice, but why just the *Green Shaft?* Anyway, as they are my favourite peas, they had to be re-sown; it was a good job I had enough left over from the first sowing.

Both the *Rocket* and *Nadine* potatoes are through the surface; despite planting at different times they are both at about the same stage. We have had quite severe frosts on two or three occasions over the past week, bad enough to damage the new shoots on many hardy shrubs, so I earthed them up to cover the tops, to be on the safe side, even though they are protected with Enviromesh.

The crops to be sown were rhubarb chard, beetroot *Mondella*, carrots *Mokum* and *Bertan*, spring onions *Santa Clause*, radishes and lettuce. In addition I decided I would sow the parsnips *Gladiator* as the spring is sufficiently advanced to ensure good germination. All seeds were sown in ½in deep drills taken out with a draw hoe, which were watered well before sowing, then filled in with dry soil. This is an effective and economical way of applying water at sowing time, as only the areas requiring moisture receive it, and none is lost through evaporation or capillary action. I always sow very thinly to avoid having to thin the seedlings before they are large enough to use as vegetables in their own right. Not only does the seed go further this way, but pest damage is also reduced, as the crops are subjected to minimal handling and therefore do not give off scents which adver-tise their presence.

Although I dislike watering unless absolutely necessary, the crops already above ground were beginning to show signs of needing a drink, so I watered the rows with a solution of Maxicrop to counter-act any stress they may have been starting to feel. Maxicrop is a sea-weed-based product which, although not a fertilizer as such, is an excellent aid to growth and is ideal for use in situations such as these.

It was almost dark before I had finished. I was cold, tired and still suffering from the effects of the 'bug' and the temptation to leave the Enviromesh sheets off was great. However, I knew that there would be nothing left by morning had I done so, so the whole plot was snugly put to bed once more.

Planting brassicas (right): holes are filled with water, allowed to drain, then planted, to ensure a good crop

Friday 18

The summer cauliflowers *Elby* are quite large enough to plant out. I plant all my brassicas in trenches. As the plants grow, the trench is gradually filled in and where the stem of the plant is buried, extra roots form, making for very firm planting. This anchorage is necessary if wind-rock is to be prevented in an exposed area such as ours. Deep planting also helps the roots to remain moist, even in dry spells. Wind-rock and drought are the two greatest reasons for failure of brassica crops, causing a variety of problems such as 'blown' sprouts, loose-headed cabbages, premature running to seed, and small, open cauliflower heads.

The roots of the young plants are inserted in holes made with a dibber in the bottom of the trench to prevent soil disturbance. The soil should remain as firm as possible at planting time; the strong roots of the brassicas will soon penetrate it, especially when previously raised in pots. I also tend to bury about 2in of the stem for

Cauliflower Elby (left)

48

extra stability. The holes were filled with water and allowed to drain before planting, and the whole trench was well watered afterwards before the crop protection sheets were replaced. These precautions should prevent any check to growth and ensure a prizewinning crop of caulies in the fullness of time.

Sunday 20

I had a visit today from two friends from Hampstead – Ruth, with whom I have worked on several books in the past, and her husband, Don. They are keen allotment holders and last year they invited me to judge the allotments on their site – a difficult job, as the best were very good, although I have to admit that many plots showed distinct room for improvement. I managed to offload a dozen Brussels sprout plants, a dozen young cabbages and six cauliflowers, which has made some much-needed room in the greenhouse. I took great delight in showing Ruth and Don our 'allotment'. The last time Ruth saw it, when we had just bought the field six years ago, it was head-high in fat hen and couch grass, and there was no shelter from the winds from Siberia. She could hardly believe how the plot had progressed, and sometimes, when I look at photographs taken around that time, neither can I.

Monday 21

Time for sowing under glass again, this time the broccoli, sweetcorn and climbing beans. The greenhouse is now at exploding point. I could make space by moving the boxes of leeks outdoors, but we are due to go away again shortly, and they will be better off remaining inside under the automatic watering until I can be here to keep an eye on them. The courgettes are coming through well, albeit more slowly than expected. This is probably due to the greatly fluctuating temperatures we have experienced over the last two weeks, especially the nights, which have tended to be extremely chilly.

Sunday 27

It rained all day on Friday, which was greeted with delight by everyone round here in the business of cultivation, as the drought situation was becoming serious. The only snag is, the weeds have sprung

Sowing sweet-corn Honey. Bantam (left) *and runner beans* (right)

to life, and as I leave for five days in Scotland later today, there is nothing I can do about it until my return. I was particularly keen to have both gardens looking good, as two very old friends are visiting us next weekend. We met back in the Sixties when I was still at school, Ron was farm manager for Burghley Estates in Stamford, and my mother was assistant accountant for the same organization. Helga has always been a keen and accomplished gardener and I gleaned much knowledge from her when my attraction to gardening started to stir. A great deal of water has flowed under the bridge since those days, but there has always been a certain amount of friendly rivalry between Helga and myself on the gardening front, although in recent years we have not seen as much of each other as we used to do. They have not seen our field and I am anxious to impress Helga with my vegetable and fruit growing prowess, and Ron with our field, as I am sure he will find much to interest him there. I just hope everything is not overrun with weeds before next Saturday.

However, I decided that, whatever else did not get done, my peas must be staked before making tracks towards the north. When we coppiced the shelter belt, we saved 40 straight, sturdy poles removed from the willow bushes and stood them outside to dry. I reckoned that by the time they were needed, they would have dried off sufficiently for them not to regrow when pushed into the soil. I was therefore somewhat surprised when I inspected the bundle about a fortnight ago and found every stake shooting from virtually every bud. In the circumstances, it would clearly have been unwise to use them for pea supports as they would probably have grown more healthily than the crop they were intended to support, so they were immediately put through the shredder and added to the rest of the mulch under the trees. At this point I felt it would be prudent to invest in something more substantial and durable which would see me right for a good many years, and obtained 40 6ft (1.8m) Metpost garden stakes, which have a steel core coated with plastic and are virtually indestructible.

Apart from the second sowing of *Hurst Green Shaft* peas, which again have yet to appear, the rest of the crop is a good 6in (15cm) high, and will need something to climb up shortly if it is not to fall over. In previous years, I have tried growing shorter varieties of peas without staking, but they eventually flop onto the ground and many of the peas rot through being in contact with the soil. This year I am ensuring this does not happen. I inserted the stakes about 2ft (60cm) into the soil, making them very rigid, and joined them together with further canes run horizontally between them and tied on with plastic cable ties, providing a very solid structure. The whole arrangement was re-covered with Enviromesh, draped over the stakes to form a permeable tent for the crop. I hope to be pleasantly surprised at the peas' progress on my return.

Early staking ensures a good crop of runner beans

May

Friday 2

Scotland was beautiful, and the trip was useful in as much as it made me realize that not all of the United Kingdom suffers from low rainfall. The drive was sunny and warm, but as we arrived the rain descended and continued to seriously pelt down for the next two and a half days. In fact, we were informed by Helensborough Horticultural Society, during my evening's appearance at their monthly meeting, that where we were staying was actually wetter than everywhere else around it. I suppose this is the price to pay for wanting a view of the loch! We managed one gorgeously sunny, warm day yesterday when the Scottish gardens glowed and we discovered that there were, after all, tops to the hills and mountains, before leaving for home this morning in a sultry heatwave.

Arriving home in the early evening, I found that the vegetable garden soil was still reasonably damp – apparently there had been rain earlier in the week – and all the crops sown on 17 April were through: carrots, beetroot, parsnips, radishes, lettuce and rhubarb chard, as well as the resown *Hurst Green Shaft* peas. The weeds were not as bad as I had anticipated, although the area earmarked for courgettes is covered with a fine crop of 'volunteer' potatoes – so much for John assuring me that when he lifted the crop last autumn he had removed 'every single one'.

Back at home things had been busy in the greenhouse. Tomatoes, broccoli and sweetcorn were well up and the climbing beans were

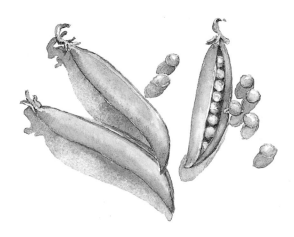

beginning to break through the surface. Every sweetcorn seed has germinated – 21 in all, so we should be in for a good crop.

After a ten-hour drive from Scotland, it would have been nice to have put one's feet up and relaxed for the rest of the evening, but Helga is coming tomorrow, and the garden round the cottage needed work on it, resulting in an evening meal at 11pm.

Saturday 3

This is perhaps the most important day of the year in our local calendar. Warm sunshine and cloudless skies are essential today and, as luck would have it, this year we were not disappointed. Spring Bank Holiday Saturday is the day of the Spalding Flower Parade, when literally hundreds of thousands of people flock to the area to watch the brightly decorated floats negotiate a three-mile route through the streets, and enjoy the stalls and other entertainment provided in and around the town. Originally known as the 'Tulip Parade', it arose during the heyday of the bulb-growing industry in south Lincolnshire, when millions of heads were removed from the vast tulip crop in order to build up the bulbs prior to lifting. The discarded heads were used to decorate floats and the parade became a major showcase for British-grown bulbs. The visiting coaches were routed through the gaudy tulip fields, where a breathtakingly colourful sight assailed the eyes – even if one did not care for this kind of thing, it was admittedly most impressive. Sadly, the fenland tulip industry is no more – most are imported from Holland and the floats are mainly decorated with crêpe paper and tulip heads sent over from the Netherlands, where they have been kept in cold storage because of the earlier springs. Nevertheless, the traditional parade and its attendant razzmatazz remain, now based at Springfields Display Gardens.

Helga and Ron were accompanying us to the Parade, which they had not seen before, despite next year being its 40th anniversary. They arrived early, so there was plenty of time to admire my labours, both in the cottage garden and the veg patch, although I think Ron, being farming-orientated, was more interested in our tenant's field of wheat and my coppicing in the shelter belt. The ornamental garden looked neat after yesterday evening's toil, and the kitchen garden appeared quite respectable, although I apologized profusely for the newly emerged crop of weeds and 'rogue' potatoes. As usual, the parade went off with its expected carnival atmosphere. It was a great afternoon as Ron, Helga, John and I have renewed an old friendship, and I picked up four – yes, *four* – superb sweaters for £10 from a stall in the arena where the floats eventually come to rest – definitely the highlight of my day.

Thursday 8

Possibly one of the things which makes living in this country so interesting is the eccentricity of the British weather. Last week we were basking in scorching sunshine, then Sunday the rain returned, and by Monday night the continuous deluge had sleet in it. Weather forecasts all week have warned of icy roads, thunder, snow and flooding, and they have not been far wrong, although in this area the extremes, as usual, have not been quite as great. Despite the temptation to stay indoors for a third day running, the state of the greenhouse demanded that urgent action be taken, if I was not to lose everything through overcrowding.

The first thing was to remove everything that did not need to be kept indoors any longer. This included the remainder of the cauliflower plants and the sprout and cabbage plants which were surplus to requirements and were collected by Ruth and David later. The Brussels sprouts are more than ready to be planted out, despite my scheduling this job for the second half of the month, and it is time to plant the cabbages, so this will be the job for the weekend. The brassica plants have not been hardened off, but the Enviromesh should prevent any problems once they are outside.

Their removal, and a complete reorganization of the bedding plants, newly planted hanging baskets, and a large container of summer bedding I take round with me when doing spring talks to gardening clubs, meant that there was enough room to prick out the broccoli and tomatoes. Twenty-eight out of thirty-two *Sungold* seeds have germinated, enough for a nice row of cordons eventually. The number of broccoli seedlings was so vast that I decided to prick them out into seed trays rather than individual pots – there were 46 *Caravel*, 36 *Green Comet*, and after pricking out 72 each of *Rudolph* and *Romanesco*, I called it a day and used the rest as a sprouting seed garnish on the trout en papillote we had for supper. Young brassica seedlings make a very good salad garnish; they taste very similar

Tomatoes
Sungold

to the so-called 'mustard and cress' obtainable in small punnets from the greengrocer and supermarket (actually rape, which is also a brassica), and are very nutritious. I can't bear to waste good vegetables, even at this stage!

Sunday 11

The weather has warmed up a little, and the showers have become infrequent enough for me to spend a day on the vegetable patch. Weeds were the main priority. Hand-weeding intricate rows of seedlings such as beetroot, carrots and parsnips is the only efficient way of keeping the crop clean at an early stage. Some people might find this job deadly dull; personally, I find great satisfaction in the neat rows of young vegetables at the end of the operation. Once the rows themselves are weeded, a light hoeing of the ground between will keep it clear of young seedling weeds. In fact, this year the weeds are less trouble than they have been since I first started growing vegetables on this piece of land, a sure sign that good cultivation over the years is beginning to pay off.

Last year's parsley is now running to seed, and although it still looks healthy, the time has come to replace it. The young parsley seedlings looked a tad small after the old crop, but they will soon romp away now it is in the open ground. It will be some time before the parsley will be large enough to pick, and I have frozen a large bag of the old leaves to tide me over. Frozen parsley has just as good a flavour as fresh, and it has the advantage of not needing to be chopped, as it crumbles easily if crushed immediately it is removed from the freezer, before it begins to thaw.

The 'volunteer' potatoes needed different treatment; each plant had to be excavated to get at the offending tuber. There is no point in merely removing the tops as the potato will continue to regrow throughout the summer. There were enough 'ground-keepers' to have kept us in potatoes for a week. I did not, however, feel this to be unnecessary waste, as we still have about half a hundredweight of the wretched things in sacks. They have sprouts on them up to 6in (15cm) long, but they still taste all right, and must be used until I can start lifting this season's *Rocket*.

The cabbages and *Icarus* sprouts were planted in exactly the same way as the cauliflowers, at the bottom of well-watered trenches using a dibber. I must confess I am somewhat uneasy about the identity of the cabbages, as when John sowed the tomatoes in the greenhouse he moved some of the packs about – they were not all labelled, as I had a 'system'. My 'system' has been destroyed and I have had to guess at the varieties. All I know is that I have four packs each containing six of each variety – or, at least, I hope so, as one cabbage seedling looks very much the same as another, though the colour and shape of the leaves do differ a little.

The climbing beans in the greenhouse are well through the compost, so I thought I should get the poles erected ready for planting out the seedlings at the end of the month. I was so impressed with the rigidity of the 6ft (1.8m) Metpost poles supporting the peas that I bought a further 40 8ft (2.4m) ones. These have been put together in pairs about 1ft apart as a narrow tent running the length of the row, joined with cable ties, and with horizontal canes along the top where each two canes meet and cross over for extra strength. About 18in are pushed firmly into the ground, so there should be little chance of the structure blowing or falling over, even when fully covered with plants.

Sunday 19

The first picking of the new season's crop is always a thrill and today I was able to pull a few radishes, albeit small ones, but very mild and juicy. I also made a start on thinning the lettuces. The leaves are just over an inch long, and large enough to add to other ingredients of a mixed salad, so it would appear that summer is well and truly on its way. *Runner bean Red Rum*

I noticed the first flower buds developing on the broad beans, so I removed the covers to allow insects access for pollination once they open. Rather than replace the covers later, I shall probably remove the tops when enough flower buds have been produced to prevent infestation by blackfly, a serious problem with broad beans.

Wednesday 21

We really are having ideal growing weather – warm and damp, with lots of showers between brilliant sunshine. Next weekend I have to make an appearance at Bicton Park in Devon, which necessitates being away from home for three days. As the sweetcorn, climbing beans and courgettes are well-grown and the month is well-advanced, I felt they would benefit from planting out before I left.

I was delighted with the *F1 Honey Bantam Bicolour* sweetcorn. There was 100 per cent germination, 21 plants in all, and every one looked strong and healthy. They were planted in their designated position, 18in (45cm) apart with about 15in (37cm) between the two rows, to form a small block rather than a single line, which will improve pollination between the plants.

The climbing beans were also very satisfactory, and, again, germination was high, with only about five failures in all, to provide 13 *Red Rum*, 26 *Desiree*, 12 *Romano* and 19 *Lady Di*, a total of 70 plants in all. The discrepancy between quantities of plants of each variety was due to there being vastly differing numbers of seeds in each packet. This does not worry me, as they should all be equally delicious, and there will be more than enough of each kind to satisfy the tastebuds. Every seedling was strong and healthy, and I could not bring myself to cull nearly half of them to leave one plant per cane, so I decided to plant two to a cane in most cases, one either side to provide about 8in (20cm) between plants. Purists would throw up their hands in horror, but I have planted more than one bean per pole in the past, with excellent results.

The courgettes *De Nice à Fruit Rond*, *F1 Gold Rush*, *F1 Greyzini* and *F1 Sardane* were less abundant, with 13 failures, but as this still gave me 20 plants – roughly five of each variety – I was well satisfied, especially as there were considerably more seeds originally than the 25 that the catalogue said were provided.

I am sure the problem with germination was because of the low temperatures we experienced shortly after I planted the seed – 65°F (18°C) is essential for rapid, healthy germination, and there have

been several periods when the temperature in the greenhouse has stayed at around 60°F (15°C) for days. On balance, however, the cost of losing a few plants is far less than resorting to putting heat on in the greenhouse, especially as if all the seedlings grow once planted out, there is still likely to be a glut! Three or four plants are usually more than adequate for the average family.

These were planted in the headland at the back of the main vegetable plot. To conserve moisture, which is essential for the formation of good and plentiful courgettes, and prevent having to weed that area in the early stages of development, I planted through Permealay ground cover fabric, by cutting a small cross in the sheeting where each plant was to go, and inserting the young courgette into the soil through the resultant hole. The plants were spaced 2ft (60cm) apart, and should fill up that area nicely when they are mature. They were temporarily covered with Enviromesh netting in case of any late cold spells, but this must be removed before the flowers start to form or lack of insect pollination will prevent cropping.

The soil is warm enough to germinate French beans, and by the end of the day, these were firmly installed.

Thursday 22

As usual, I am late planting up my hanging baskets. It is safe to put these outside around the second week of June, and I like them to be well established before I hang them up. As there are over 20 of them, space in the greenhouse can be a problem, so everything which can be moved outside must go. The removal of the beans, courgettes and sweetcorn has created a little room, but this still leaves two big boxes of leeks which will not be large enough to plant out for a little while yet. However, they will come to no harm if they are introduced to the great outdoors, so they are now sitting on the garden table – I hope one of the cats does not adopt this as an interesting bed, which has happened with other trays of seedlings in the past!

The broccoli seedlings are developing more slowly than I expected, so they will need indoor protection for a short time yet.

Friday 23

The early peas are covered with flowers, and this season promises to be a bumper one. If I were to be put on the spot and asked to name my favourite vegetable, it would have to be fresh peas – home grown, of course, followed closely by baby broad beans; and it looks as though I shall not be disappointed here, either, as there is an abundance of flower, much in demand by bees, on the broad bean plants.

Monday 26

At last, my first green salad of the season. I removed about 8oz (227g) of well developed thinnings, which will be enough for about three meals when mixed with young Good King Henry leaves and given a good vinaigrette dressing.

Friday 30

The bitterly cold easterly winds are beginning to have an effect on the crops. Early carrot germination has been extremely patchy, the *Santa Clause* spring onions are virtually non-existent, and *Radar* are the only French beans showing any signs of an appearance. Each day we hope for an end to the drought, but watering will become a must if rain does not fall soon.

June

The garden in early June

Tuesday 3

Weatherwise, we seemed to have slipped back two months. The cold, windy weather we should have had in late March is making life most unpleasant, but the vegetables are soldiering on, as are the weeds, so a thorough hoeing and hand-weeding was the first priority. Despite the lack of rain and the position of the potatoes, in the driest part of the plot near the older shelter belt, the tops are very strong and healthy, and promise a good crop of earlies in about a month's time. The turnips and lettuces needed thinning and this yielded 8oz (227g) baby turnips, 2lbs (900g) turnip tops and 3lbs (1.4kilo) lettuce leaves. The baby turnips are delicious eaten raw like radishes, but I decided to cook them for lunch; they only required bringing to the boil and then serving tossed in butter (or sunflower spread if you prefer) and chopped fresh parsley.

There is a limit to the amount of lettuce which can be consumed in salads, and here my culinary ingenuity starts to come into its own, as I refuse to discard anything which has eating potential. Lettuce is quite tasty when shredded and stir-fried in hot oil for a minute or two, then seasoned with a little stir-fry seasoning which can easily be obtained from the local supermarket – you need quite a lot for a decent serving so it is a pleasant way of disposing of a large amount of more mature thinnings. My great stand-by at this time of year is one of three excellent lettuce soup recipes I have used for many years. These all freeze well and are perhaps even nicer during the depths of winter than at this time of year.

Lettuce soup (1)

Serves 4

Ingredients
2 oz (50 g) butter or margarine
1 lb (450 g) lettuce leaves
4 oz (110 g) spring onions
1 level tablespoonful flour
1 pt (600 ml) chicken or vegetable stock
$\frac{1}{4}$ pt (150 ml) milk
Salt and pepper to taste

Method
Sauté the roughly shredded lettuce and chopped spring onions in the melted butter or margarine until soft. Add the flour and cook for another minute, then gradually blend in the stock. Cover, bring to the boil and simmer for about half an hour, then remove from the heat and allow to cool. Pass through a sieve or blend thoroughly in a food processor then add the milk and seasoning and reheat gently – do not allow to boil.

Lettuce soup (2)

Serves 4

Ingredients

2 oz (50 g) butter or margarine

1 lb (450 g) lettuce leaves

1 medium onion

1 medium potato

$\frac{3}{4}$ pt (450 ml) chicken or vegetable stock

Salt and pepper to taste

4 tablespoons double cream

Parsley

Method

Shred lettuce and dice onion and potato. Fry gently in a large saucepan in the butter or margarine until soft but do not brown. Add stock and milk and bring to the boil, stirring all the time. Cover and simmer gently for about 10 minutes, then remove from heat and cool. Liquidize or pass through a sieve and then reheat to serving temperature. To serve, swirl on a tablespoon of cream per bowl and garnish with roughly chopped parsley.

Lettuce soup (3)

Serves 4

Ingredients

1 lb (450 g) lettuce leaves

1 large onion or 1 large bunch spring onions

1 clove garlic

Generous $\frac{1}{4}$ teaspoonful ground nutmeg

1 tablespoon lemon juice

$1\frac{3}{4}$ pints (1 litre 50 ml) chicken or vegetable stock

Salt and pepper to taste

To serve – $\frac{1}{4}$ pt (250 ml) single cream, chopped parsley

Method

Fry the onion and garlic for 5 minutes (2 minutes for spring onions) in a little butter, margarine or oil. Add the shredded lettuce leaves and fry for a further two minutes. Add the stock and other ingredients and simmer for 10 minutes. Liquidize or sieve and reheat. Allow to cool slightly then serve with a liberal swirl of single cream in each bowl and coarsely chopped parsley.

John is no problem in helping to clear the lettuce mountain, but my inventiveness is taxed to its limits when it comes to turnip tops. I love them, shredded and steamed for about 8 minutes then served with freshly ground salt, pepper, nutmeg and a knob of butter. They have a unique flavour which can only be described as a cross between nettles, spring greens and spinach – sounds awful, but I can assure you it isn't! However, I know when John eats a whole portion of something on his plate first that he wants rid of it before devouring what he considers to be the more palatable ingredients of his meal (he was obviously brought up well as a child) and this can happen when turnip tops are on the menu. As a result I disguise them heavily by straining off any excess liquid, beating in plenty of double cream and gently reheating, then somehow they assume for him the status of an acceptable vegetable.

Because of the height of the vegetables, I decided to risk leaving off the Enviromesh sheeting for the time being, as there should be little chance of insect, bird or cat damage doing serious harm to the crops. The peas and beans have grown much taller than usual, and required tying up as, even with a framework of stakes, they were flopping everywhere. The runner bean plants are halfway up their canes and flower buds are already starting to appear low down on one or two plants. The cold weather late last month has taken its toll of the

Beans (right) *and peas* (far right) *mid-June*

courgettes, with the loss of two plants, but there will still be ample for our needs.

The leeks in their boxes are now the thickness of a pencil, so the last job of the day (thank goodness for long, light evenings!) was to plant them out in their cropping positions. There were 125 in all, and these were planted in two rows with 6in between plants and 9in between the rows. The plants were removed from the trays, the surplus compost shaken off and the tops trimmed, before dropping into 6in deep holes made with a dibber. The holes were filled with water to firm the plants but not backfilled with soil – this happens automatically as the rows are cultivated.

Planting leeks in holes filled with water

Wednesday 4

The only good thing about a spell of poor weather is that it comes as a pleasant surprise when it improves. It was cold and misty when we drove over to Stamford to the dentist's; by the time we emerged again the sun had broken through, it was several degrees warmer and the bitter wind of the last few days had dropped. I really cannot understand why people dread a trip to the dentist – for me, it is the only time I get an opportunity to lie still and relax for a quarter of an hour, other than when I am in bed at night, so I quite enjoy it. However, this might also be due to the fact that ours is a very dishy dentist, although thankfully John prefers the chairside assistant and the booking secretary. On the way home, having considered yesterday that it was unlikely that any more French beans, spring onions and *Mokum* carrots would germinate satisfactorily, I bought replacement packets for each crop.

Unfortunately, it was not possible to buy the same varieties over the counter as they were only available by mail order, so instead I opted for the safe, tried and tested varieties *Early Nantes* carrots, *White Lisbon* spring onions and *Cropper Teepee* French beans, a similar variety in habit to the sparsely growing *Purple Teepee* already *in situ* but with green pods. After lunch I resowed the gaps, and started off a successional crop of mixed radishes as the first sowing is now becoming depleted. These radishes have proved to be a most interesting selection, much remarked upon by visitors who have sampled them in multi-coloured salads, including long types like *French Breakfast*, as well as round red, pink and white forms and even the odd black-shinned one resembling *Black Spanish*, which is usually grown as a winter variety.

The broccoli plants, although not large, are ready to plant out. There were scores of the wretched things, so I selected the strongest five of each of *F1 Green Comet*, *F1 Caravel*, *Romanesco* and early purple sprouting *Rudolph*, and dumped the rest onto our long-suffering friends Ruth and David, whose smile was a little fixed when they looked in the bags and discovered the quantity. I pointed out that if there just happened to be too many, surely they must also have friends who would be delighted with some free plants which are at present selling at enormous prices in local garden centres! It always seems odd that people are prepared to pay about three times as much for a lettuce or brassica seedling for growing on, than it would cost them for the mature plant ready for consumption from the greengrocer.

I toyed with the idea of replacing the Enviromesh on the part of the garden where the broccoli is sited, as they would make a most desirable meal for marauding pheasants, pigeons and the brace of partridges which has recently appeared. However, as the row is sur-rounded by well-established brassicas, I decided against it and hoped the baby plants would not be spotted.

The first flower truss has appeared on the *Sungold* tomato plants, so I knew they were ready for planting out. There were 28 in all – four I gave to our next-door neighbours to raise in growing bags, the rest were planted in a single row in the vegetable patch with about 1ft (30cm) between plants, and provided with a cane each for sup-port. They will be grown as cordons, with the side shoots removed as they appear. Although I have found it is possible to grow *Sungold* as bushes, the resultant plants are large and unwieldy, with many small trusses of smaller fruit, whereas as cordons the plants are more attractive, with better fruit which is easier to pick.

There is still no sign of rain, and the forecast is not optimistic as far as the end of the drought is concerned, so I felt I could withold water no longer, and had the sprinkler on for a couple of hours. On checking, the water had penetrated well, and with luck this should be the only irrigation needed for some time, even if rain does not fall. It is far better to water thoroughly once than sprinkle regularly, which generally does more harm than good to the developing crops, as their roots are discouraged from delving down to find moisture.

Tuesday 10

Well, wouldn't you believe it? As soon as I decide to water, the weather changes. Despite the forecast, last Friday we had thundery rain all night, followed by a showery Saturday, followed by thunder on and off all Sunday! The weather was greeted with delight by everyone, despite a spoilt weekend for many, as the anxiety about the increasing shortage of domestic water is now a real issue.

Today is drier, and tomorrow I leave for the NEC and four days of BBC *Gardeners' World Live*, so I spent the day making sure every emergent weed was safely on the compost heap, then yet another nourishing bowl of lettuce soup – number three recipe this time, as we had chicken over the weekend and it seemed a good way of utilizing a large bowl of stock from the carcass. As you will have realized by now, things would have to be serious before we starved!

Wednesday 11

We drove to Birmingham in torrential rain, which lasted most of the day. *Gardeners' World Live* was as interesting as ever, and the weather did not dampen the spirits of the visitors, but if I have one criticism, it was the attitude of many of the stand holders, both in the floral marquee and in the main halls, where most of the trade stands were situated. I bought an interesting device from Wolf Tools – a gadget with a removable handle which could be used with my range of snap-on handles. This consisted of a miniature push-pull hoe which could be turned over to provide a tiny 3-pronged cultivator, just the job for working between individual plants and narrow rows in the vegetable garden. Unfortunately, the salesman, who was clearly afflicted with a severe attitude problem, nearly made me resist the purchase. The same distinctly unfriendly approach was exhibited by two of the four nurserymen from whom I bought plants, and at the end of the first day I was left wondering just why some firms want to exhibit at shows such as *Gardeners' World Live* if they find the whole thing so distasteful. How off-putting such encounters must be to the many new gardeners who attend flower shows in order to gain encouragement and experience.

Sunday 15

The show was blessed with two reasonably fine days, followed by a wet Saturday which severely curtailed my outdoor performances on the BBC Radio 2 stand. I returned home yesterday evening, hoping to get a good day's work in on the garden, only to be rained off completely. I felt extremely sorry for everyone involved in the last day of *Gardeners' World Live*, as it was apparently extremely damp there.

During the morning, the torrent eased off sufficiently for me to pick six lovely, white, tennis-ball sized *Tokyo Cross* turnips and the very first peapods of the season, which yielded just under 4oz (114g) of succulent *Daybreak* peas after shelling.

As we are now running out of last year's old potatoes and I am

loathe to buy new ones until my own earlies are ready, these turnips, which will mature very quickly from now on, will provide a good substitute. Unlike more conventional varieties, they require very little cooking and are best eaten when still slightly crisp. I cut them into dice or thin sticks and steam them for about 10–12 minutes, before tossing them in butter and finely chopped parsley. My cookery books list innumerable ways of cooking and serving peas, but I still just prefer to cook them for about two minutes in a small amount of boiling, salted water with a sprig of mint. In the unlikely event of a glut, I will happily eat a whole dish of young garden peas cooked this way for supper, but as I like to freeze as many as possible for winter use, these occasions are very few and far between.

Monday 16

There were plenty more showers today, so it seemed sensible to find jobs indoors, including the sowing of the winter cauliflowers *Walcheren Winter Pilgrim* in a seed tray in the greenhouse. I had intended to sow these outdoors in a seedbed in a spare bit of ground at the back of the vegetable patch, but when I opened the packet, there were so few seeds that they would probably have got lost if they had not received individual treatment. The quantity is no problem; there will still be more than I shall need. Later in the day, the weather settled down a bit, so I had an hour's weeding among the crops. The sudden wet weather has made all the radishes run to seed, so there was nothing I could do but pull them up and feed one of my compost bins with them. The rain has also made several of the onions and shallots run to seed. This is probably divine retribution as, driving past a field of onions a few days ago, I felt very smug to see so many with flower heads when mine were, at that time, so perfect. Onions and shallots which have run to seed will not store well, but are still all right for use throughout the summer, providing the flower heads are removed immediately.

Tuesday 17

Between yet more showers it was mini-harvest time chez Daffers. Tomorrow I have to work at the Lincolnshire Show with BBC Radio Lincolnshire, doing various outside broadcasts from the flower tent and stage performances near the Radio Lincolnshire stand. We have farming friends living adjacent to the showground, and it is more convenient to site our caravan in their farmyard overnight and walk to the show, rather than risk getting bogged down in traffic on the morning itself. Like most farmers, Bill does not grow vegetables except for large quantities of limited kinds, like potatoes, for selling wholesale, so he and Jenny are delighted with any surpluses from our garden. Tomorrow evening we have been invited to supper by our friends Keith and Carol Fair, who run the well-known Valley Clematis Nursery near Louth in north Lincolnshire. Keith and Carol are at present winding down the nursery and Carol is converting the areas previously used for propagating and sales into a private garden which, she hopes, will eventually make them self-supporting in fruit and vegetables. In the meantime, they, too, are grateful for any donations. After Lincoln we travel to Ely to meet former colleagues of John's for a meal at The Lamb, so today I have to gather enough for Jenny and Bill, Keith and Carol and ourselves to see us through until Sunday evening.

First, there were the forty – yes, forty! – golf-ball sized turnips to be divided between us. Jenny loves peas, so I removed all that were ready on the *Daybreak*, which almost filled a plastic supermarket bag. The yield from a 6ft (1.8m) row of *Daybreak* so far is incredible. The rhubarb chard is ready for picking, although the leaves are still quite small. Rhubarb chard is cooked in exactly the same way as spinach when young, and tastes very similar. I prefer to grow chard or spinach beet to true spinach, as it does not run to seed so easily and has a much longer cropping season. It is recommended only to remove a few leaves from each plant, so as not to weaken them but our soil is so fertile that I generally cut the whole row back in order to remain on top of what usually turns out to be a copious crop. Almost four pounds was produced on this occasion from a mere half row, so you can see what a profitable plant rhubarb chard is. Carol is very fond of spinach-type vegetables, so she should be pleased with her half-share.

The lettuces are hearting up well, and it is now possible to pick ones with quite decent hearts, so I pulled nine – three heads each. They will remain fresh for quite some time if gathered with the whole root and some soil and stored in a polythene bag in a cool place. There is a good selection of varieties in this mixture; the packet lists *All the Year Round, Cobham Green, Little Gem, Lobjoits Green Cos, May King, Saladin, Valmaine Cos* and *Winter Density*. I have to say that the picture on the packet is somewhat misleading, however, as it shows a red butterhead type and a frisée, looseleaf variety which could have been very useful for decorative green salads, and the picture in the catalogue shows a deep red looseleaf form as well, whereas my row has no red lettuce at all. As I particularly like red lettuce for garnishes, I will probably sow a short row in the next week or so.

Friday 20

It has hardly stopped raining at all since Tuesday, but our local water company still insists there is a water shortage as we are apparently having 'the wrong kind of rain' which is not reaching the aquafers! Consequently all the bore-holes are running low. Both days of the Lincolnshire Show were damp in the extreme, although Wednesday's heavy showers let up long enough to allow my *al fresco* question and answer session to go ahead.

The bag of peas I took to Jenny's yielded almost 8oz (227g) when shelled, so she was delighted. Jenny is the nursing sister at our local surgery, and our friendship started when she was assigned to look after John three years ago on his discharge from hospital, following complications after an operation for prostate cancer. Unbeknown to us at the time, she had just lost her own husband; looking back on it the brave face she put on things was astounding. A year later she met farmer Bill, and the two are now a happy item, but although Jenny spends most of her time at Lincoln, she has still retained her surgery job. Jenny knows where my vegetable patch is situated, and is not very honourable where peas on the vine are concerned, so it is not unknown for her to drive in and help herself when we are not around. For this reason, I have found it better to 'buy her off' with a regular bagful; in this way I can keep an eye on the way the crop is performing. In all honesty, I also have to say that there is absolutely no point in her even thinking about pilfering peas, because there are none ready for picking left!

I was most impressed with the progress Carol has made in reland-scaping the nursery. She already has a very aesthetic-looking vegetable plot, although it is behind mine as she started later and she is further north which, although we live in the same county, does have an effect on maturing dates. She was thrilled with the bag of rhubarb chard; she likes to serve it steamed, with pepper and salt and a knob of butter, topped with a poached egg (or two!). She is looking forward to the time when the broad beans which are her second favourite vegetable do not come frozen out of a packet, and that time is happily fast approaching.

On the way from Lincoln to Ely I stopped off in the pouring rain at the vegetable patch to pick up a further supply for the weekend. Bill shot some rabbits yesterday evening; it was my job to skin and dress them this morning. Although I popped most of them in the freezer, I saved a couple for a rabbit stew with turnips for Sunday lunch. The tops will provide greens, and a touch of luxury will be supplied by the first *Oregon Sugar Pod* peas which are now being produced thick and fast. I picked what turned out to be 8oz (227g) before becoming totally saturated. There was also another small boiling of peas which I shall probably use cold to liven up a lettuce salad over the weekend.

Turnip tops have a unique flavour when steamed

Monday 23

After a weekend of thunder, heavy rain and high winds the excessively tall broad beans had suffered, and in spite of being supported by canes, they had fallen over. I should have taken the tops out earlier, but there was little sign of blackfly so the job had been missed. Up to 18 inches (45cm) at the top of each plant bore no pods, so I decided to remove all this surplus growth to let the plants concentrate their energy into producing good beans. Some of the shorter shoots were young and tender enough to eat – broad bean tops can be cooked like spring greens and have a texture like spinach, with a flavour similar to the beans themselves.

Many of the climbing beans have reached the top of their poles and there are a few red flowers on the plants already, so it promises to be an early crop. All the courgette plants have flower buds on them, and two are sporting open flowers, unfortunately all male, but this is quite usual at the beginning of the season and female flowers

should be produced in an adequate quantity once the plants have settled down to a flowering routine.

The resown carrots, French beans and spring onions are well up, so there is no germination problem now the ground is adequately moist. I was hoping to be able to pull a few of the new sowing of radishes today, but although the tops are well developed, the roots have yet to swell.

Leaving off the Enviromesh does not seem to have affected the crops unduly, except that the partridges have had a go at the first brassica plant of each sort nearest the gravel – obviously easy pickings, as none of the others have been affected, but the damage is only superficial so far.

Tuesday 24

John is my second husband. My first, Tony, who has not remarried, lives about thirty miles away. It is nice that, after a prolonged period of 'readjustment', John, Tony and I are good friends. John helps Tony with DIY and I plant up his hanging baskets, in exchange for which

Tony invites us for the occasional meal. I suppose it is inevitable that this usually turns out to be my vegetables and my cooking, while the men talk about things electrical (John is an audio engineer and Tony an electrical engineer so they have discovered they have interests in common). Today was delivery day for Tony's summer baskets, and we arrived well laden with vegetables for the evening meal and to see Tony through the rest of the week. We took him a rather magnificent iceberg-type lettuce, some mature rhubarb chard, the inevitable turnips and tops, and a bag of mixed mangetout peas *Oregon Sugar Pod*. I also included the first *Sugar Snap*, which look exactly like normal peas, but the pods are tender and fleshy, with lots of little juicy peas inside, so the whole thing is cooked. Of the two mangetout-type peas I am growing this year, I think I prefer the *Sugar Snap*, which have a more pronounced pea flavour. Unless the pods are very young, both types require 'topping and tailing' before cooking, to remove the stringy bits running down each side of the pod, rather like the older varieties of runner bean. However, providing this is done, even quite old pods are tender and juicy after boiling or steaming for only a few minutes, so that some crispness of texture is retained.

Rhubarb chard is another excellent dual-purpose vegetable; once the leaves are large enough, the red mid-rib can be removed and

Rhubarb chard

76

The garden in late June

cooked and served like asparagus. I find it best to steam it for about ten minutes so that the attractive colour is preserved; it is then served with freshly milled salt and pepper and a generous knob of butter (garlic butter is even tastier if you like garlic). The leaves minus the mid-ribs can be cooked like spinach.

We had an interesting meal by candlelight during a power cut, the present husband, the ex-husband and I. Watching the lightning illuminating the church on the hill at Collyweston, where I spent my formative and embryonic gardening years, I could not help reflecting on the way my life has evolved.

Wednesday 25

Back to the drawing board today, with site visits to two new houses on a very attractive small development near Grantham. One garden is in excess of a quarter of an acre, the other one around half an acre, and yet neither client was interested in growing vegetables or fruit. Perhaps they are too young and upwardly mobile.

Sunday 29

John is a very clever chap, and can turn his hand to virtually anything if necessary, and this is quite necessary at present. Made redundant by the BBC about eight years ago, he has to rely on relief work at local radio stations to keep the wolf from the door and retain his self-esteem. As there are hundreds of BBC engineers in a similar position, all competing for freelance work, these engagements are few and far between. His precision DIY work has achieved an acclaimed reputation within our circle of friends, with the result that he is asked to tackle jobs very far removed from the profession for which he is qualified. The latest bizarre commission is to tile a farmhouse kitchen and hall floor for Jenny and Bill with Fired Earth slates; these give the effect of instant antiquity but are the very devil to lay. Logistically, it is better for John to stay in Lincoln while he does the job. It works well if I stay with him in our caravan as I can feed him properly, so he does not exist on a diet of cut-price baked beans, and I can also bring work with me which I find difficult to settle to at home, where I am surrounded by a hundred and one other distractions. I have three garden layouts to complete, and John's tiling expertise will give me an opportunity to be a captive designer.

So tonight we travel to Lincoln again and I must stock up with veg for the week, and the communal meals we will no doubt share. The mangetout peas are in full bearing now; I picked 4lbs (1.8kilo) of *Sugar Snap* and nearly 6lb (2.7kilo) of *Oregon Sugar Pod*. Bill has just wound down his beef herd, so there is a freezer-full of prime cuts, no doubt some turnips will be useful and lettuce always comes in handy.

Last Friday I attended the Anglian Flower and Garden Show at Wimpole Hall near Cambridge. It was a quagmire, and I had to admire the gardening stalwarts (some in high heels!) who turned up in their droves, although it has to be admitted that at the last question and answer session the three people who comprised the audience had individual attention. One of my editors on *Garden News* had taken himself off to the Glastonbury Festival, and, by all accounts, we in East Anglia should stop moaning. Our vegetable garden is remarkably healthy, although water has started to stand between the rows where I have walked in order to pick crops. I corrected this by breaking up the compacted soil to about four inches with my three-pronged cultivator, after which the water drained away, proving that there is no serious compaction of the land.

July

Tuesday 1

Another day of sunshine and showers, and another new garden to measure. Diana and Michael are friends and clients of many years' standing. Recently they have renovated and extended an old corn mill and the garden is now ready for landscaping. They have a totally separate, self-contained area near the kitchen which will make an ideal vegetable plot. They are lucky in this, as so many people have to make do with a section of the garden which is less than ideal, either aesthetically or cultivation-wise, for this purpose. In between downpours, we measured up and discussed the potential of the rest of the land before returning to our temporary establishment in Lincoln – John back to his tiles and me to the ever-active word processor and drawing board.

Late this evening John decided to go for a walk on Bill's farm, which adjoins the Lincolnshire Showground. On the way home in the dark, a fawn greyhound attached itself to him. It was thin, scruffy, and obviously very hungry, but would not allow itself to be touched. It would, however, eat as much food as we were prepared to offer: first the remains of the farm collie's supper, then a tin of steak and onions which I always keep in the caravan in case we require an instant chilli con carne or shepherd's pie. Neither the police nor the local RSPCA was interested, we retired to bed, on the logical assumption that the dog would either still be around in the morning, or would have disappeared.

Wednesday 2

The greyhound is obviously no fool, as this morning he was still there, circling round and eyeing us from a safe distance. It was a situation which clearly required further action and eventually we managed to obtain the telephone number of the local dog warden. She arrived, full of confidence and a bag of biscuits. The dog viewed the biscuits from afar, but was not prepared to accept them from her hand, although it would eat them off the ground before retreating to

a safe distance once more. After several hours of fruitless persuasion, the farm foreman suggested that if he fetched his wife, plus their tasty young labrador bitch, the greyhound, who had obviously been a fine specimen of a young male before living rough, might be distracted enough for the warden to slip a rope over his head. He did indeed find the young lady delightful, so much so that he decided to give in without a fight and allowed himself to be lifted into the back of the van, on the possible assumption that if he went off with the dog warden, there might be more of the same sort of thing at the other end of the journey.

I asked what was likely to become of him, and the warden told us that he would be kept in rescue kennels for a statutory week, then if he were not reclaimed, he would be rehomed or put to sleep. She thought that both reclaiming and rehoming were extremely unlikely, however, as this particular area was a favourite dumping ground for unwanted greyhounds – he was the fourth that week – and the local dogs' homes were saturated with them. In a moment of compassion I gave her our name and telephone number, and watched them disappear out of the farmyard and into the unknown.

Saturday 5

The weather has improved at last, and we arrived home from Lincoln in brilliant, hot sunshine. The unsettled period has had no bad effect on the vegetables; all crops have put on a pleasing amount of growth in our absence, especially the tomato plants, which I had to tie in to the canes immediately. They had also produced a lot of side shoots, and as *Sungold* are much better grown as cordons, they all had to be pinched out immediately, before they got too big and spoilt the parent plant. This made my fingers delightfully yellow and fragrant, but I confess I love the smell of tomato plants on my hands, and the stains will wear off eventually, so I am not bothered.

The broad beans were now much taller than the temporary supports I had provided for them. There was a lot of unproductive growth at the top of the plants, and I could cut off at least 2ft (60cm) without removing any flowers or forming pods. This has made the plants much firmer, and will encourage them to make big, juicy beans.

Monday 7

A day for the hoe, as it has been gloriously hot all day, and the severed weeds soon wither in such weather. Hurrah! the first of the *Express* beans are ready, and, after shelling, yielded 8oz (227g) in all. Broad beans always seem rather wasteful to me, as there is much more to discard than there is to eat, so I often pick the immature beans and cook them whole. However, the nitrogen-rich pods make good compost-heap material. On our patch we have a table and two seats made out of huge poplar trunks we originally obtained for firewood some years ago, but which, given a touch of 'sculpting' with a chain saw, have turned out to be much more useful. When the weather is like this I usually take a few minutes to sit here and shell my peas and beans as soon as they are picked, so the residue can be added to the compost heap immediately. I have to confess that if I take them home to shell and I am in a hurry, it is very tempting to stuff them down the waste disposal unit in the kitchen. The original gadget was installed before we moved to the cottage – whether I would have one if I were to design a kitchen from scratch is arguable, but it has to be admitted that it is a useful piece of machinery for disposing of waste food (not that there is much in our house!). Our local refuse collection more rightly belongs to the Middle Ages and consists of delicate, black plastic bin bags (dustbins and other animal and insect-proof containers will not be emptied) which are left at the side of the road for the best part of a day every Wednesday for removal. It is therefore important that anything likely to putrify is disposed of by some other means. While uncooked vegetable waste can be composted to advantage, meat waste and

cooked vegetable remains are unsuitable for rotting down, so are most hygienically disposed of by flushing down the sink via the waste disposal unit. My excuse for disposing of the odd raw vegetable parings in a similar way is that our local sewage sludge is supplied to farmers for spreading on the fields, and if my garden does not benefit, at least the parings have not been wasted in the greater scheme of things.

Raw vegetable waste makes excellent compost

The *Sugar Snap* and *Oregon Sugar Pod* peas have been busy in my absence, and have produced a further 1lb (454g). We are just about managing to keep up with the peas at the moment, but I will need to start freezing some very shortly.

The *Rocket* first early potatoes are in full flower; once these begin to fade I shall have an exploratory dig to see if my first boiling of new potatoes is ready.

Tuesday 8

A scorching day, just right for a trip to rural Warwickshire where I am to perform my annual task as one of the judges of the GIMA awards. Members of the Garden Industry Manufacturers' Association are invited annually to submit products to be judged in various categories by a panel comprised of people involved in different aspects of the gardening industry. It is always an interesting day's work and gives me an opportunity to see the best of what is on offer to the amateur gardener, although it is also increasingly arduous as more and more products are entered each year. The venue is chosen to be as convenient for the judges as possible, and this time Warwickshire seemed the most appropriate, though living where we do, it is always something of a trek to reach any destination which is easily accessible to the rest of the judges.

We were nearing the end of our three-hour trip when the mobile phone rang. It was the owner of the rescue kennels, informing us that, as expected, nobody had claimed the greyhound, and were we interested in giving him a home? Now, until I embarked on my somewhat itinerant way of life as a writer and broadcaster, I had always had dogs around and enjoyed the special relationship and companionship they provided. However, when the last one died some fourteen years ago, I felt that it would be far too difficult to spend long periods away from home and have any animals other than cats, which are thankfully independent. John has never owned a dog and, as far as I could see, is not really keen on them anyway. But, amazingly, he seemed to strike up a relationship with the hapless greyhound during his encounter with him on Lincoln Cliff, and so, when the phone call came, he did not raise the string of objections I expected him to. In recent years I have spent more time at home, with the caravan as a second abode, so owning a dog is not the problem it was a year or two ago. Clearly a decision had to be made, so I agreed to collect the miserable hound later this week. Well, at least there will be another mouth to eat our vegetable surpluses!

Wednesday 9

Today was one of my regular 'specials' on Radio 2's *Jimmy Young* show. As most of the staff working on the programme live in London and work long hours, vegetable growing is out of the question for them, so it is nice to be able to take a 'goody bag' of fresh produce with me when I go to Broadcasting House. The box of lettuces, radishes, turnips, rhubarb chard and mangetout peas was eagerly seized upon when I arrived. London was unbearably hot, and it was a delight to be able to escape to the country once more after my stint on the radio.

Thursday 10

Yesterday's heatwave turned overnight into a dull, cool day today, rather a let-down but eminently suitable for doing some work on the carrots, which attract carrot fly far less if they are handled only minimally, and then on a dull, still day. Patches of the *Mokum* early carrots have developed quite well after all, so I thinned these carefully and managed to gather a decent bunch of 'finger' carrots, of a flavour so delicious that it seems a shame to cook them. I shall cut some of them into julienne strips to use in salads; the rest will only require steaming for a minute or two as a hot vegetable. The re-sown *Early Nantes* carrots have germinated and grown well to fill the gaps left by poor germination of the *Mokum*. I thinned these a little, as I seem to have sown them rather thickly and the roots will not fill out properly otherwise.

This afternoon we travelled to the north of the county to collect the dog. We were surprised to see how much better he had started to look in the space of just over a week, having been de-flead, de-ticked and fed some decent food. We had to give an instant name for the kennel's records – we had not given this much thought as we felt something would come to us once we got to know him. However, faced with the question, I said the first thing to come into my head, which was Carlton, after the village near which he was found, so Carlton, or Carl, for short, I suppose it will have to be. On the way home, we obtained the largest bed we could find because we were informed he liked to sleep fully stretched out, a large sack of dog meal, several tins of tripe as he had apparently been fed this while he was in the kennels, a hide chew in the shape of a knuckle bone, a smart greyhound collar and an extending lead. The bill came to over a hundred pounds and I have a feeling that our life is going to change considerably from now on.

Friday 11

We are both very tired today. We are also wondering just what we have done in offering a home to a dog we know nothing about. The garage is very snug and comfortable, so we installed his bed, plus several blankets, in there and, around eleven o'clock last night put him, and his chew, to bed. At one in the morning I was awoken by a terrible banging and howling, a noise destined to raise the dead, to say nothing of the neighbours, and which was obviously emanating from the garage. On unlocking the door, the first thing to greet me was a large pile of wood shavings; this had been produced by Carl's methodical gnawing through the substantial, framed, ledged and braced door. The garage floor was covered in unmentionable deposits and most of the gardening tools I keep at home had been strewn across the floor and dunked in this mess. The cosy bed was unused, the chew untouched.

For the sake of the neighbours, the emergency option was to put the horrible dog to bed in our room and sort the disaster out in the morning. Once his enormous 'pie-dish' bed was relocated on our bedroom floor, the wretched animal slept the sleep of the just, which is more than we did.

Carl walks well on a lead and looks as though he may well turn out to be a kind and gentle dog, given time, but there is much work to be done, which may well end in divorce, although, surprisingly, John has so far reacted better to his misdemeanours than I have. He probably thinks all dogs behave this way, but in all my years of canine experience, I have never come across anything quite like the scene which met my eyes in the middle of last night. In an attempt to tire him out before bedtime, we decided to introduce him to our field and the vegetable patch. It is obvious that he has been trained to run, and he was soon racing at tremendous speed, all grinning teeth and bulging eyes, along the track between our tenant's wheat and the eastern shelter belt. The crops are maturing thick and fast. I returned home with a bag of lettuce, 4oz (114g) shelled weight of *Daybreak* peas, which are about the last of the early ones, a sack of both sorts of mangetout peas and another of *Express* broad beans. I removed all the edible leaves on the rhubarb chard and then discovered that the turnips, some of which had grown very large after all the earlier wet weather, had been attacked by cabbage root fly. Several were also starting to run to seed, so I felt it was best to harvest all the remaining roots and freeze them, as within another week most of them would be inedible. The cabbage root fly is possibly my fault, for not re-covering the plot with Enviromesh, but the damage is not bad and the uncovered crops have been much easier to tend and harvest.

Most of the tomatoes have now reached the top of their canes, which are about 4ft (1.2m) tall. There is no point in letting them grow any taller than this, as it will only encourage them to make fruits which will never ripen, at the expense of the trusses lower down. I am therefore removing the growing tips as they get to the top and at the same time all the side shoots which have been produced since the last pinching out. It seems that side shoots are growing as soon as my back is turned, and it is essential to check the plants every day or two for more. Any that are missed soon produce long, thick side branches and within a very short time, the whole row can become a terrible muddle if the job is skimped.

Saturday 12

It is very satisfying, seeing the empty drawers of the freezers gradually filling up with new season's produce. As spaces begin to appear in the deep freezers in the spring, I fill them with temporary food such as bread and cakes which can be quickly used when room is needed for more permanent provender. We have three freezers, two large ones, and a smaller one for convenience built into the kitchen units. It may seem surprising in view of the fact that there are only two of us, but these are always full, especially from this time of year onwards, though mercifully not entirely with vegetables! Today I froze 5½lbs (2½ kilo) *Express* broad beans, which is about the last of this variety, 3lb (1.3 kilo) rhubarb chard, 1lb (454g) *Oregon Sugar Pod* and 2lb (nearly a kilo) *Sugar Snap* peas. There was quite a lot of waste on the *Tokyo Cross* turnips but I managed to rescue 5lbs (2¼ kilo) for freezing, which should be ample for winter soups, stews and casseroles. Now the whole of that crop has been harvested, I estimate that the row has produced about 12lb (5½ kilo) in all, not including the tops which were used as a green vegetable, which is a very worthwhile amount, although not unbeatable.

Sunday 13

Some time ago, a retirement home in our local market town asked me to redesign their garden following building alterations. The job is now well under way and this afternoon I have been invited to attend their fête, inspect the work and draw their raffle. True to form, the day dawned thundery, with sultry sunshine interspersed with heavy showers. This morning, in order to give Carl the opportunity to chase, I spent a couple of hours in the vegetable garden, digging over the area where the turnips had been, and removing yellowing lower leaves from the brassica plants, which can harbour pests and diseases if not cleared up regularly. During this time, Carl vanished, eventually to return covered in blood. Thinking he had had some unpleasant accident or, at the very least, had an encounter with the attractively white-stemmed but exceedingly vicious Rubus cockburnianus in the shelter belt, I cleaned him up tenderly with warm water and TCP – the only antiseptic I had to hand down there – and made him a soft bed in the shed from old potato sacks and horticultural fleece. We were later informed by our next-door neighbour that he had, in fact, consumed a large hare on her doorstep and the blood, therefore, was that of the unfortunate rodent and not his own. Sympathy dwindled rapidly thereafter.

The weather was decent enough to settle down for the fête. The new garden is developing slowly but surely. There will eventually be an area where the residents can, if they want to, grow vegetables and, although the yields are not likely to make much difference to the shopping bill of the retirement home, it will, at least, provide those taking part with the satisfaction of supplying the kitchen with fresh, home-grown produce. The goal is also to sell any surpluses at open days such as the one I have just attended, which will help to swell the coffers of the Amenities Fund. It would be nice to think that similar establishments might embark on similar schemes in the future – there are many elderly people who, although not confident enough to continue living alone, still want the pleasure and gentle exercise which gardening provides.

Monday 14

The unsettled weather has caused many of the lettuces to run to seed. This evening, a warm, balmy one, I removed the worst of these and consigned them to the compost heap to allow the remainder to develop properly. Lettuces which have just started to seed are still quite edible; however, as they shoot upwards, the leaves become tough and much stronger-tasting. Although they can still be used in soup, we are, frankly, sick of lettuce soup so, not having a tortoise or rabbits, the compost heap was the only answer. I re-sowed the now sizeable gaps with more lettuce to provide a successional crop when the first sowing has been used.

I turned my back for a minute, and another load of side shoots had appeared on the tomato plants, so these had to be pinched out without delay. Now the main growing tips have been removed, the side shoots just below the tops grow very rapidly and need to be checked almost daily for their appearance. Plenty of long, strong trusses are forming, and the fruits on the earlier ones are already swelling, so if we get some warm sunshine from now on, it should not be too long before we are able to pick the first tomatoes of this summer.

Some of the runner bean plants have reached the tops of their poles. Rather than stop them at this point, I twisted them along the cross poles running between the uprights. The whole thing will end up a terrible tangle, but it will in no way affect the performance of the beans, and an enormous crop looks likely. There are flower buds low down on the *Red Rum* runner beans, so it will only be a short time before the first boiling will be ready. To anyone enjoying growing vegetables, this must truly be the most exciting time of the year.

Wednesday 16

It rained solidly all day yesterday, so a fine, sunny morning today was most welcome. I had a sudden urge for a taste of new potatoes, so took Carl to the Patch for an experimental dig.

The dog is settling well, given his background, but has a propensity to stray if left on his own in the vegetable garden when somebody is not amusing him. For this reason, I decided to tie him to a long piece of polypropylene rope which we happened to have in the shed. It was really thick and substantial – the type used for towing gliders. It was a relief to think I could give my undivided attention to the garden for an hour or so without constantly having to be on the lookout for the latest of Carl's misdemeanours.

It was with dismay, therefore, that less than five minutes later, I looked up to find him standing beside me, a short piece of rope attached to his collar, the rest lying on the ground near the shed, still tied to the ground anchor I had screwed firmly into the soil. Next job will be a trip into town, I fear, to purchase a long, strong metal chain.

The potatoes produced by the first *Rocket* seed tubers were quite adequate for two meals for the two of us – just under 1lb (454g) of well-shaped, round tubers. Most were quite large, up to 3in (7.5cm) long, so the actual number was not great. Nevertheless, the thrill of inserting my trusty potato fork into the soil and unearthing fresh, young tubers was just as great as every other year I have grown my own – one of the loveliest sensations, to my mind, of the vegetable gardening year.

The ways of cooking potatoes are legion, but there is no better method of dealing with the first serving of the year than to wash them gently to remove the soil, cut in half if necessary, and steam with fresh mint sprigs until just cooked. New potatoes can cook quickly, and steaming avoids their boiling into the water if the cook's attention is distracted when they are nearly ready for serving. It is not necessary to remove the skin from new potatoes; in fact, if the potatoes do not have a residue of chemicals on their skins, it is not necessary to peel them at all, even when 'old'. The highest concentration of vitamins is in and just below the skin, so in peeling them, you are throwing the best part away. I always cook more than I need for immediate use as there is nothing better than home-grown new potatoes in a potato salad. My special potato salad consists of diced, just-cooked potatoes which are still firm to slightly crisp, mixed with mayonnaise, finely chopped chives or spring onion tops, one or two hard-boiled eggs depending on the amount made, and enough cayenne pepper to give it a bit of a 'bite'. This is almost a meal in itself.

At last I am able to pull spring onions. There are very few *Santa Clause*; those which are ready are juicy, red tinted and highly attractive, but the yield is disappointing. Thompson and Morgan, the suppliers, tell me this variety is going to be withdrawn from their catalogue next year because of its problems, and advise me that *White Lisbon* is still the most reliable spring onion for general use, which is just as well, as it is *White Lisbon* with which I filled the gaps when I re-sowed and it is mainly these I am pulling now.

Sunday 20

When the weather is hot in July, it is very, very hot. We returned this afternoon from a weekend spent on the outskirts of the Derbyshire Dales, attending a Saturday evening National Trust music and fireworks concert in the grounds of one of their properties. The crops were bursting at the seams so, regardless of all the other things needing to be done after being away from home, it was essential to gather all the vegetables which would spoil if left another day. The first of the *Hurst Green Shaft* peas were ready, there were yet another 3lbs (1.3 kilo) *Sugar Snap* peas, a generous picking of maincrop *Imperial Green Longpod* beans, just over 2lb (1 kilo) of the first sowing of dwarf French beans (mainly *Aramis* and *Vilbel*), a sackful of rhubarb chard and no fewer than nine out of the fifteen *F1 Elby* cauliflowers, which were all a good size at around 8in (20cm) in diameter. The hot weather today has been so tiring, however, that when we arrived back at the cottage we felt like doing little except sitting out in the garden into the muggy twilight – the harvested vegetables must wait until tomorrow for attention.

July 21

Eight out of the nine cauliflowers had to be frozen. The curd is so dense that one will last us at least three days, even if it is also added raw to mixed salads. I prefer to eat cauliflower lightly steamed, without any trimmings, though I do sometimes, in deference to John, serve it with cheese sauce. We are now able to have at least two vegetables other than potatoes with our main meals. This week it will be cauliflower and the 'filet' type French beans, which are much too young and tender to waste by freezing them. They can be cooked

Hurst Green
Shaft *peas*

whole at this size, just removing the tops and tails for appearance's sake. There are also the *Hurst Green Shaft* peas and, as a change, I shall cook some of the rhubarb chard, although I picked far too much yesterday afternoon for our current needs. I therefore offered the rest to a neighbour; she had not encountered this vegetable before but was fascinated by its appearance, especially the colour. I am loath to give away any mangetout peas yet as I know they will be very welcome in the depths of winter, so the latest picking of these was also frozen.

Tuesday 22

The weeds in our vegetable garden are nothing like as bad as in previous years, but this is only to be expected, as the area and its surroundings have been kept much neater than when we first bought the field. There is no doubt that the adage 'one year's seeding, seven years' weeding is true; the more you keep on top of the garden, the less likely it is to get on top of you. I must admit I really enjoy hoeing, and as there were a few seedlings beginning to appear between the rows this morning, I took action while they were still young and chopped them off. When the weeds are at this stage, it only takes about two hours to hoe the whole vegetable plot. In previous years the job has occasionally got the upper hand and it has taken days to clean up the crops, as weeds soon grow too large to be hoed off and have to be removed individually by hand. When weeds get to that size, the crops suffer, as they have to compete for moisture, food, light and air with the interlopers.

I was ready for another root of potatoes. The second digging yielded slightly more than the first, but it would appear that each *Rocket* tuber is going to produce plus or minus 1lb (454g). This should be sufficient to see us through to the maincrop, providing we go carefully. Serving a number of other vegetables with each meal ensures we do not have to fill up on potatoes, as can be the case when produce is in short supply.

There are a number of baby *Mondella* beetroot which can be pulled to add variety to our inevitable daily salads. The roots are only just over 1 inch (2.5cm) in diameter, but require boiling for only a short time and have an incredibly sweet flavour. At this stage I usually serve them hot with a plain white sauce, or cold without pickling in vinegar. Cooked beetroot does not last more than a few days without becoming mouldy, even in the refrigerator, so it is advisable only to pick enough at a time for immediate use. The leaves would look, cook and taste very much like rhubarb chard if I felt inclined to save them, but as I have a surfeit of this type of vegetable at the moment, I considered the needs of the compost heap were greater than mine on this occasion.

Wednesday 23

The dog has an upset stomach. It must be all the vegetables we are feeding him. Generally speaking, I am advised, cooked vegetables added to a dog's diet are a good thing, but greyhounds, apparently, have a rather delicate constitution, so I shall stick to his proprietary all-in-one meal in future.

Thursday 24

This evening, a local horticultural society visited both gardens. It has thundered most of the day, and at one stage I was in two minds whether to cancel the visit, but miraculously the weather improved for a short time after tea and remained fine just long enough for the members to see as much as they wanted to. The orchard and vegetable garden at the Patch provide a good contrast to the cottagey, ornamental garden around the house, so there is usually plenty to see. Having these occasional public inspections does wonders for keeping one on one's toes in the garden. The next big event is in just over a fortnight, when both gardens are open for BBC Radio Lincolnshire's 'Go for Gold' Appeal in aid of an annually nominated local charity. From now onwards, therefore, all effort in the gardens will be concentrated on producing a showpiece worth travelling miles to visit, which is quite some feat.

Friday 25

The second sowing of lettuce is big enough to thin out and use in salads. I was pleased to see that this time there are going to be some red plants amongst them. Most of the old lettuces have now run to seed, so I removed the rest of them to allow the new ones to fill out. I also made a late sowing of *Little Gem* lettuce which should supply us with salads nearly until Christmas if the weather is not too harsh.

The crops of onions and shallots look really splendid, especially the onions, which are all large, uniform and turning a lovely, golden brown colour. One or two of them have run to seed; these I removed for immediate use rather than waste them. Onions which have run to seed have thick, sappy necks and a fleshy stalk up the middle which begin to rot in a short time if you attempt to store them. On the other hand, the flavour is not affected at all, and if there are too many for the present time, 'bolted' onions and shallots can be blanched and frozen (see Appendix 1) if they are double-wrapped to prevent the smell from affecting the rest of the freezer.

Sunday 27

I shall be away from home for several days next week, first in Manchester working on my programme for Granada Satellite Television, then in Daventry where I am to present the prizes at the GIMA Awards Dinner. As the weather was fine and dry, I again harvested as much as was ready in the vegetable garden. I would love to be in a position to pick only what I need for the next couple of meals just before I need it, as the quality and food value is highest in just-harvested crops. Unfortunately, my lifestyle – and the fact that the vegetable garden is three miles from home, making the logistics of harvesting more complicated – just will not allow this. Indeed, how many people in our modern world would, in fact, be able to do this? On the other hand, home-grown vegetables picked and preserved immediately will still be more nutritious than those lying about on the shop shelves, so I must be thankful for this, at least.

The first runner beans were ready – only 4oz (114g) on the *Red Rum* plants, but enough for a tasty serving for two, and there were another two servings provided by the French beans. All of the original four varieties have some beans on them, but the bulk of the crop will come when the re-sown *Cropper Teepee* mature. I was also able to pick the first head of *Green Comet* broccoli which will be enough for two large portions. After shelling, there were 3½ lb (1.6 kilo) *Imperial*

Green Longpod broad beans. As I shall not need these until after my wanderings next weekend, I froze them temporarily without blanching – they will be used long before they start to deteriorate in flavour. The *Hurst Green Shaft* peas are also swelling thick and fast, so I treated a further 12oz (342g) shelled weight to a similar temporary treatment to the beans.

There are tassels on the sweetcorn already. This means that, all being well, we should have an early crop of, hopefully, well-pollinated cobs, although this will depend on how much breeze we get in the next weeks. Despite planting in a block, if the weather is still, damp and cool when the plants are starting to mature, many of the kernels end up unpollinated.

From now onwards, it is essential to check the base of cordon tomato plants as well as further up the stems for unwanted shoots. On my inspection today I found several quite well-developed shoots had appeared from just above ground level and were threatening to turn into extra plants. They had to be removed with secateurs as they were too thick to be pinched out. Seeing those lovely stems already bearing trusses of flowers lying on the ground seemed a dreadful shame, but sometimes in gardening you have to be cruel to be kind.

The first fruits are developing on all four varieties of courgette plants. They are, as yet, very small, but I may not be able to visit the Patch again for another week, by which time they could have turned into small marrows, so I removed them and ate them, stir-fried with a further picking of *Sugar Snap*, in butter, garlic and pepper and salt, with my supper.

The *Perfect Ball* cabbages are, indeed, perfect balls, very hard and well-formed, but capable of putting on more growth. As there are plenty of other vegetables available at present, I felt there was no

need to start on the cabbages yet. They will stand for several weeks *July vegetables* and will make a pleasant change later in the summer. I have to be careful not to remove too many vegetables like these before the open day; the visitors do like to see clean, well-filled rows burgeoning with produce.

Three more cauliflowers were ready for cutting and freezing, the two left will be ready for eating fresh on my return.

August

Friday 1

BBC Radio Lincolnshire's Go For Gold open day at our gardens is imminent, so everything will revolve round that for the next ten days. The garden at home is really no problem as it is heavily planted, and there is virtually no weeding involved, so work there just consists of lawn cutting, dead-heading, and making sure that one plant is not crowding out another. Feeding and watering of hanging baskets, window boxes and other containers is automatic, thus saving valuable time that can be more usefully spent on the vegetables.

Today has been sunny and dry, ideal for hoeing, and by the end of the afternoon the rows of produce were immaculate. The over-sized side shoots I removed from the tomatoes last week, but rather untidily threw down on the ground instead of feeding the compost heap with them, had actually rooted where they touched the earth – a great testament to the toughness of tomatoes and the quality of our soil.

Monday 4

Visitors to a vegetable garden like to see a plot burgeoning with food and as the *Walcheren Winter Pilgrim* cauliflower plants are now well-grown I decided to plant them out so they will have settled down before the open day. With a heavily cropped piece of ground such as this, it is sometimes difficult to find room for successional vegetables such as winter-hardy cauliflowers if a full range of varieties is to be cultivated from the outset. The only available space was the row vacated by the turnips. In a perfect scheme, this would not be a good choice, as I do not like replacing one brassica crop with another for fear of a build-up of pests and diseases, especially club root. Once club root has got hold of a piece of ground, it is difficult to grow good brassica crops for many years afterwards, although there are ways round this, such as growing each plant in a comparatively large container full of good, clean compost so that it is semi-mature before it is introduced to the infected soil. It is then much more capable of developing into a good plant, despite still possibly contracting the disease, which causes swollen, malfunctioning roots resembling dahlia tubers.

Club root is a fungal disease which is generally introduced by bringing infected plant material into the garden, and sometimes it arrives on footwear if someone has walked on land infected with club root beforehand. It is less of a problem in alkaline soils. The vegetable garden was limed at the beginning of the season, so the pH is still high, and as to my knowledge there is no way I can accidentally introduce the disease, I am prepared to risk replacing the turnips with the cauliflowers.

The other snag with planting the cauliflowers in this spot is that they will not all be harvested until the middle of May next year. By this time I shall be on year two of my crop rotation, and the new season's brassicas will all be on that part of the garden where I grew potatoes this year, so it will rather upset my overall plan, but the second year planting will just have to take place around it.

Tuesday 5

I had a good dummy run for Sunday's great day this afternoon with a visit from the local Townswomen's Guild – actually the members plus their husbands. One of the features about the Patch they found most interesting was the shelter belts, and the excellent microclimate they had produced in the vegetable garden and orchard areas of our field – almost tropical on this August afternoon. The willows, dogwood and other subjects which were coppiced after Christmas have already made up to 8ft (2.4m) growth, creating a private and tranquil environment in which to work and rest. Trees and bushes are an unusual sight in the south Lincolnshire fens, where every square inch of ground is cropped, regardless of whether the produce is actually needed or not. If market prices are not favourable, crops are often ploughed in or fed to sheep, to be immediately replanted with the same or similar; not much crop rotation here. Copses, hedgerows and headlands take up land which can be otherwise exploited, so our shelter belt is something of a rare sight in the area. I often wonder whether, in fact, commercial yields would be the same, or maybe even higher, if shelter belts were more often provided on farms in our area. At least they would prevent the terrifying and appalling 'blows' which frequently occur in dry springs in the light soil areas to the west and south of us. It is not unknown for the whole of a newly seeded crop to disappear in a cloud of dust during a high wind, where hedgerows and trees have long been grubbed up to create huge fields for the machinery which is now part of the farmer's obligatory equipment. Grant aid is now available for the replacement of this very necessary part of crop husbandry; trees and hedges do not grow overnight and it will be many years before any beneficial effect is felt on a large scale. I doubt if I shall notice any appreciable change in our local landscape in my lifetime, more's the pity.

Saturday 9

Phew, wot a scorcher! as they say. Burning hot sunshine from dawn to dusk, which is what I did not really need if the gardens are to look at their best tomorrow. By evening both the cottage garden and the vegetables were looking limp and jaded, and as I did not want to start watering at this stage, I began to pray for a heavy dew.

To make my open garden days more attractive to a wider range of visitors, we borrow next door's drive, which is flat and wide, for a row of stalls. These vary from year to year, but usually offer wrought iron, cakes, an ice cream barrow, plants from local garden centres and nurseries, and the inevitable bric-à-brac and tombola. The conservatory is turned into a lace shop, and the garage becomes the tea room and raffle display. BBC Radio Lincolnshire are excellent caterers, as they do a great deal of this kind of thing for charity, and take over our thankfully well-equipped utility room at the rear of the garage for their food and drink preparation.

Organizing something like this is an ideal way of getting John to tidy out the garage. Nearly everything is taken down to the shed at the Patch for the event, the floor is painted with easy-clean concrete paint, and for some time afterwards the garage is immaculate. Then the clutter starts to reappear until by the end of the year we are just

about back to square one. The shed, being the receptacle of everything we do not want to dispose of but which is in the way at home,
never gets tidied and therefore gets progressively worse.

This year we have so much surplus produce that we are also able
to have a vegetable stall. We were therefore engaged until the small
hours in making saleable packages of a huge pile of produce brought
home on the trailer. All the runner and French beans are now bearing prolifically and needed weighing out and bagging in 1lb (454g)
lots. The mangetout peas required similar treatment, but the
rhubarb chard is bulky and I thought it would sell better in ½lb (227g)
packets. I felt I could spare three *Perfect Ball* cabbages and some
heads of *Green Comet* broccoli, which we wrapped professionally in
Cling-film to keep them fresh, and there are now enough *White
Lisbon* spring onions to make into attractive bundles. For the sake of
effect, I also added a few bunches of *Bertan* carrots, which are quite
large enough now for 'finger' carrots, and *Mondella* beet, as the
orange and red colours make the display more striking when distrib-
uted amongst the green packets of beans and brassicas. I topped off
the stall with some bunches of parsley and other herbs which I grow
in the cottage garden – fresh herbs always sell well.

Much as I would have liked to offer a wider range of produce,
there are certain crops I really could not spare if we are to have
enough for our own needs. Onions, shallots, *Hurst Green Shaft* peas,
Mokum and *Early Nantes* carrots and potatoes will have to be
admired but not sampled by the visitors.

Saturday 10

There is always a fear that an open day will be a wash-out, and there
have been times in the past when the thunder has rumbled and we
have cast our eyes upwards anxiously for most of the day. Today,
thankfully, was not one of them, and if I have one criticism of the
weather, it is that is has been *too* hot. Profits were down, mainly
because of the excess heat, and even the refreshments did not vanish
with the normal gusto, although, as usual, the cakes and home-
grown vegetables disappeared rapidly.

Garden lovers are, in general, super people, and the atmosphere is
always good at these functions. There is never any damage to the
plants or garden itself, nor the tiniest bit of litter. To prevent the pilfering of cuttings, I allow anyone who wants them to have what he or
she wants, within reason, providing I am asked first, the material is
suitable, and I supervise the cutting being taken. I can then advise on

how it should be treated and provide some damp material to convey it home in. A donation in the collecting box is also most acceptable, but not obligatory.

Whatever the profit, there is a tremendous amount of satisfaction at the end of an open day when everything has gone as intended. There is a great deal of forward planning involved in this kind of function – highways departments, police, planners to inform, the press, radio and other media to contact, helpers to recruit. When the gates close at 6pm I invariably say that this will definitely be the last time ever, but the pleasant memories long outlast all the stress and hassle of the moment.

Monday 11

One of the difficulties of having two gardens three miles apart is that when I open the gardens to the public, I cannot be in both places at once. Especially when BBC Radio Lincolnshire is joint organizer, it is necessary for me to stay at home, in the ornamental garden, where I can be available for interviews in the programme being broadcast live to the county and surrounding area. Visitors to the vegetable garden are therefore looked after by informed helpers who are on hand to answer questions if necessary. Because I am not present, the visitors feel they are more able to make audible comments about what they are seeing, and while they are mostly favourable, the odd criticism *is* overheard and relayed back to me in due course. On this particular occasion it was the onions which apparently caused concern amongst the more experienced vegetable growers. The hot weather has ripened them off rapidly over the last week or so; the tops have bent over and are now drying off, and the bulbs could have been lifted last week, but I wanted them to be on display in their growing positions. Keen-eyed visitors spotted this and made remarks to the effect that they would have had them out of the ground long before this. Not being present at the time, I was unable to justify myself and explain that so would I have done if it were not for the Go For Gold day. There is no excuse now not to go ahead and do it, especially as the forecast is for fine, settled weather for the next week or so.

When onions are ripening, it is sometimes necessary to lift them partly out of the ground with a fork, to detach the roots from the soil and therefore speed the drying-off process. It is also often recommended that if the tops are particularly thick and lush, the bulbs ripen better if the tops are bent over manually. However, I find that as the onions ripen, most of the tops bend over without assistance, and those that do not are likely to be poor keepers, because their necks are particularly thick and soon begin to rot in storage. It is these onions, therefore, which should be kept separate from the bulk of the crop and used first.

Today I loosened all the onions and shallots with a fork and left them lying *in situ*. The top inch or so of soil is quite dry, and the bulbs should start to dry off nicely left here for a few days. At the end of the week, I shall either remove them from the garden and spread them out over the south-facing concrete if the forecast remains good, or take them into the shed to finish off.

Wednesday 13

I picked the last of the *Imperial Green Longpod* broad beans this morning. They have produced an excellent crop and have provided plenty for the freezer. By this time of the summer, broad beans have tended to lose their novelty, but come the depths of winter they will have grown most desirable again. I am using cold, cooked broad beans a lot in salads at the moment, mixed with plenty of chopped parsley and a garlic vinaigrette made from a crushed clove of garlic, a pinch of salt and pepper, a teaspoonful caster sugar, a teaspoonful French mustard, ¼pt (150ml) olive oil and 2 tablespoons vinegar.

If space allows, I leave the spent bean haulms in the ground as long as possible to continue their job of fixing nitrogen in the soil. Sometimes there is even a second crop from new shoots produced from the base of the plants, or, if you want something different in the vegetable line in early autumn, you can pick the tips of these new shoots and steam them for a few minutes before serving them, seasoned with freshly ground salt and black pepper, with a knob of butter, or puréed with cream.

Thursday 15

The first *Sungold* tomatoes are ripe. I was able to pick around 8oz (227g), but several of these had disappeared before the bag arrived home. It is tempting to eat cherry tomatoes like *Sungold* like sweets, and I am sure they are much better for you. The plants themselves are still growing, and it is still important to remove the side shoots regularly before they take over.

I gathered what appears to be the last of the *Hurst Green Shaft* peas. Although there are still plenty of *Sugar Snap* peas and a few *Oregon Sugar Pod*, I shall miss real, round peas without shells. There are a few bags in the freezer, but I shall save these for dishes in the future where no other accompanying vegetable than fresh garden peas would be quite right. As with the broad beans, I shall leave the old pea haulms in the ground until I need to prepare it for something else. Pea and bean roots can be dug into the ground to advantage, together with their nitrogen-fixing bacteria, while the tops are a good source of nitrogen in the compost heap.

The yield from the *Sugar Snaps* has already been quite incredible – the plants are still growing and I expect them to crop at least until the end of the month. Cold mangetout peas are just as tasty as hot ones; I serve them in salads tossed in vinaigrette dressing with finely chopped apple mint added.

When I was a child, my favourite vegetable was beetroot. This was probably because it was the first one I ever grew, and I had never seen it before. I found the whole appearance totally thrilling, and therefore the flavour equally exciting. Nowadays I am not so keen. I enjoy it as a hot vegetable, with a white sauce, in moderation, and occasionally as a cold pickle, but I would not want to grow more than a short row annually. I think the problem is that until recently, when it was closed, we lived down-wind of the local sugar beet factory. The pong from this when in full operation was quite revolting, a cross between stewed soil and boiling washing, and, unfortunately, the smell of boiling beetroot is not dissimilar. We still have beetroot in the freezer from a previous year which, although it has retained its flavour, is wet and limp once thawed, although it makes a very pleasant and pretty cold soup. Today, however, I picked a most attractive bunch for supper; hot beetroot goes very well with certain baked fish dishes, especially cod, and is less earthy in flavour if served with a generous knob of butter and sprinkled with ground ginger.

Cold beetroot soup

Ingredients
Approx 1lb (454g) cooked beetroot
2 medium onions
Small clove garlic
1 medium carrot
4 oz (114g) hard cabbage (*Perfect Ball* is ideal for this)
$\frac{1}{4}$ pt (150ml) water
1pt (600ml) beef stock made with two stock cubes
Salt and freshly ground pepper to taste
1 tablespoon lemon juice
Soured cream or plain yoghurt

Method
Grate the beetroot, carrot and onions and shred the cabbage finely. Simmer the beetroot, carrot and onions in the water for 30 minutes, then add the cabbage, beef stock and lemon juice and simmer for a further 20 minutes. Liquidize and season to taste. Chill thoroughly for several hours or overnight. Swirl soured cream or yoghurt onto the surface just before serving.

Tuesday 19

Rain threatened this morning, and the onions and shallots, which have been drying off so well on the concrete outside the back door of the shed, had to be moved indoors quickly before they became wet again. I spread them in a thin layer on the dry concrete floor of the shed. They are almost ready for stringing, but some of the tops are still damp – they should be really strawy before tying together otherwise moulds will form which can affect the bulbs themselves.

Wednesday 20

My turn for the *Jimmy Young Show* has come round again, and with it another opportunity to offload some of my surpluses. It is impossible to keep up with all the French and climbing beans; there were around 5lb (2¾ kilo) and 30lb (over 13½ kilo) of these respectively. Every time I visit the vegetable patch I pick and freeze a large bagful of beans and still they are beating me. There has been no time to blanch them before freezing them, but I have found that there is no serious deterioration in flavour for at least six months if this job is not done, and everybody I know who freezes her own produce agrees with me.

In addition to the beans there was 3lb (1.3 kilo) of mangetout peas, three bunches of *Mondella* beetroot, over 2lb (1 kilo) *Sungold* tomatoes, some secondary heads of *Green Comet* broccoli which have appeared since harvesting the main heads, some young *Little Gem* lettuces which are filling out nicely, several bunches of *White Lisbon* spring onions and the last two *Perfect Ball* cabbages – it will not be long before the first of the *Bingo* are ready, much earlier than I had expected, so I do not mind having a break from cabbage for a week or two. The vegetables were conveyed to London in two large paper sacks, which I dumped in the middle of the office floor and told the staff to sort out amongst themselves. I have no doubt that the fresh food was most welcome, but did not dare to ask later what had become of it in case I was told it had been taken to the tip!

One thing is certain, a vegetable garden of the size that ours is, is more than capable of feeding the average family. Such surpluses, if there is no outlet for excess produce, could be reduced by sowing rows only half as long, but it is nice to be in a position to give others a chance to sample produce of a quality seldom experienced without a vegetable patch of one's own.

Friday 22

John's middle son, daughter-in-law and young Amy are staying with us for August Bank Holiday. I shall not want to spend all weekend in the garden, so I had a quick tidy-up this morning before spending the day cooking in anticipation of their arrival later this evening. Fortunately all three enjoy vegetables, so I should be able to impress them with the fruits of our labours so far. The *Rocket* potatoes are now finished, and today I dug the first of the *Nadine*. These have had a difficult time as they are growing very close to the shelter belt, which has taken a lot of moisture and nourishment from them. Digging was difficult as the tree roots, which were carefully removed in the spring, have invaded the area again and were well wrapped round those of the potatoes. In spite of this, the yield from the first root was reasonable, about 1½ lb (nearly 700g), although, as with the *Rocket*, there were a few very large tubers rather than many medium-sized ones, but these will be good for baking, so it is no problem.

Wednesday 27

A pleasant Bank Holiday was spent by all of us, despite the weather acting true to form, and raining most of the time. It is now time to get back to work before the garden becomes a wilderness. The courgettes, like the potatoes, are suffering from being too close to the trees, but I manage to pick enough once a week for about four servings, in addition to which the attractive creamy, dark green and yellow skins of the *Greyzini*, *Sardane* and *Gold Rush* respectively look most attractive when thinly sliced in salads, as a replacement for cucumbers, which I have not grown in the greenhouse this year. I am not sure about the usefulness of *De Nice à Fruit Rond*; its shape makes it difficult to place in the culinary scheme of things. I find it best to harvest this particular variety when it is about the size of a large golf ball and steam it lightly so it is still *al dente*, preferably with a selection of the other three varieties as well for visual interest.

The shed, actually a steel-clad farm building, becomes very hot inside when the sun shines on it, so the onions and shallots have finished drying off well. I like to make them into strings as soon as possible after this so they can be hung up in a cool, light, dry place to over-winter. The best spot for them is the garage at home, which has an even temperature. It would be unwise to leave them in the shed,

which becomes very warm in fine weather, but can freeze during cold spells.

I am delighted with the *Jet Set* onions. There were 70 bulbs which had not run to seed and have already been used. Each bulb was uniform in shape and size, and weighed an average of 8–10oz (227–284g), making a total storable crop of around 40lbs (over 18 kilo). The yield by weight of the shallots was slightly lower, but the number was enormous, as each original bulb had produced up to 18 new ones, so there were hundreds of them! As shallots are generally used in cooking by number rather than weight, I think we shall be all right for a few months.

The correct way to string onions is to plait the dry tops together, adding new bulbs as you work up the plait. However, the tops are often not long enough to make a firm string, as was the case with mine today. The alternative is to bunch three or four bulbs together, tie the tops together with raffia just above the necks, then bunch another three or four together and tie these to the tops of the first bunch, and so on. In this way, there is always a reasonably long piece of tops to work with and, when you have reached the end of the string, you can plait these together and turn them over on themselves to make a strong loop with which to hang the string on a suitable hook. The string becomes quite heavy as onions are added to it, and to prevent the whole thing dropping to pieces after a while under its own weight, it is advisable not to make it longer than about 2ft (60cm) if you use the raffia method of stringing. When the onions are suspended, there should be a free circulation of air around the string, as rotting can occur where the bulbs are in contact with another object, even themselves, so it is wise not to bunch too many together at a time.

The alternative to stringing onions for storage is to use the nets in which they are supplied to shops and this method is better than nothing. The problem here is that all the bulbs are lumped together under the pressure of their own weight. If rotting occurs in the centre of the bag, it can often be undetected until you notice a foul smell or see a dribble of revolting liquid on the floor, by which time many of the onions surrounding the rotten one have been affected as well.

Stringing onions is a time-consuming but gentle job for a pleasant late August afternoon, and it kept me occupied for several hours until the sun disappeared in a feverish ball behind the alders and it was time to return home with the fruits of my labours.

Thursday 28

John is wearing his audio engineer's hat today and is working at BBC Radio Lincolnshire, relieving a member of the engineering staff who is on holiday.

He cannot wait to get back into harness if the opportunity occurs. I suspect gardening is not really him, unless it is construction or mending something. I sent him off with a good supply of runner beans, rhubarb chard and tomatoes. He left them in the radio station kitchen with a notice on them 'Please help yourself' and reported back tonight that everything disappeared rapidly, with the exception of the rhubarb chard. I do not know whether to believe him or not about the rhubarb chard as I know he does not really like it himself, and I suspect he might be involving his colleagues in his support.

If I have a criticism of *Sungold* tomatoes, it is that the skins split very easily. They can be perfectly firm when you remove them from the plant, and as soon as you let go of them, the skin immediately bursts, rendering them useless for anything except freezing and cooking. For this reason, I usually pick them when they are very slightly under-ripe. They are still perfectly edible, in fact I think the flavour is even better as it is more tangy.

Friday 29

The *Sugar Snap* peas, which have done such sterling service throughout the summer, have now succumbed to powdery mildew. The foliage is covered with a white, powdery deposit which has damaged the youngest growth and more or less put an end to the crop. I must confess this has almost come as a relief as we really cannot cope with many more. For the sake of garden hygiene, I removed all the pea haulms this morning and made a little bonfire of them in an old dustbin on the gravel hard-standing outside the shed. Adding them to the compost heap could encourage the disease to hang around for another year. I did, however, leave the roots, together with their nodules of beneficial soil bacteria, in the ground, to be dug in at a later date. I washed the plastic-covered poles with disinfectant – it is at such times that this type of 'cane' is advantageous, as it is far more difficult to disinfect bamboo ones.

Removing the haulms revealed a lot of *Sugar Snap* pods which I had not noticed at the time. These were now very mature – too tough to eat whole, but could be shelled like ordinary peas. They made enough for a nice serving for two; they were deliciously sweet and tender and required steaming for only a couple of minutes.

The *Icarus* Brussels sprouts are developing well, and baby sprouts are already forming in the leaf axils. The bottom leaves of all brassicas tend to turn yellow and drop off as the plants mature, and it is a good idea to remove these and clear them away to prevent the build-up of pests and diseases – especially slugs and snails – which can occur if they are left lying on the soil underneath the plants. While I was doing this, I noticed with some alarm that cabbage white butterflies have already been busy on all the brassicas, and the first batch of caterpillars has hatched and is beginning to work its way through the choicest young leaves. In previous years I have kept them covered with Enviromesh at this time of year, but even this has not been entirely successful as the butterflies have managed to find the tiniest chink in the covers, got underneath, and had a positive orgy of egg-laying, undisturbed by birds and other predators. Assuming that the plants were protected, I have often not noticed what was going on until the caterpillars have chomped their way through the crop and

Brassicas (above) *and* Icarus *Brussels sprouts* (left)

done immeasurable harm. For this reason, I decided earlier this month not to take barrier action against the pests, which were almost certain to invade this area of the vegetable garden, and cross that bridge when I came to it. The bridge has now presented itself, and on further inspection, I noticed that there was also a heavy infestation of brassica whitefly which, apart from making the vegetables concerned very difficult to clean for cooking, weakens the crop and causes damage which can encourage rotting in the heads of cabbages and sprout tops. At this time of year, one thorough spraying should clear up both problems which, as the summer ends and the autumn wears on, may be sufficient to protect the crop until maturation. I used a permethrin-based insecticide; permethrin and similar chemicals are artificially produced chemicals resembling pyrethroids, which are extracted from the pyrethrum daisy. They are extremely effective but not persistent and are not absorbed by the plant, so what is achieved is a quick knock-down without any residual effect which could taint the food in question.

Organic gardeners may argue here that a more 'environmentally friendly' way of treating caterpillars is to use the bacillus which is available for applying as a drench, the idea being that the caterpillars become infected with a particularly horrible disease and die off. The problem with this is that the bacterium cannot differentiate between cabbage white caterpillars and other similar garden pests, and butterflies and moths which do no harm in the garden. As we have precious few of these anyway in the intensively farmed environment in which I live, I see no point in adding to their problems.

Saturday 30

The peas may be finished, but the climbing beans run on apace. The *Red Rum* and *Lady Di* are regularly producing large bunches of fleshy, tender, stringless pods. The white-flowered variety, *Desiree*, has not produced quite such a prolific crop. The plants were slower to mature and one or two died after planting, but this is likely to be because they were in a less favourable part of the row, which was overshadowed to a certain extent with raspberries at one side and blackberries at the end. *Romano*, the climbing French bean, has produced a good quantity of beans unlike any you are likely to see if you do not grow them yourself – flat, knobbly and long, with a flavour which is a cross between a dwarf French bean and a runner bean. I usually serve these separately, as they deserve to be sampled as a type of bean on its own, rather than combined with other beans from the same row.

The dwarf beans are not likely to produce as heavy a crop as I have achieved in previous years as the initial germination was so patchy, and the re-sown gaps have struggled because they have been dwarfed by mature plants all around them. However, there have been enough for immediate use, and for freezing the few packets I may need in the winter for recipes calling for French or filet beans. *Aramis* has been the best cropper to date, followed by *Radar*, although the weight of *Aramis* has been greater, as *Radar* is picked when no longer than 3–4in (7.5–10cm) for the best texture and flavour. *Radar* is a very useful bean for serving cold in salads, steamed for about three minutes so it is still crisp, then cooled and covered in a plain vinaigrette dressing. *The Prince* is a standard, flat-podded variety which tends to look rather old-fashioned when grown against the round-podded forms, but possibly has the best, typically dwarf bean flavour, of all. *Purple Teepee* is useful, as the purple pods are easy to see for picking, but as far as flavour and appearance is concerned, differs little from *Cropper Teepee* which was the variety used for resowing the gaps. It is claimed that they are both easier to pick than other varieties of dwarf beans as the flowers and pods are borne above the foliage, but this is not entirely the case, and the whole plant needs to be checked when picking as beans may be found underneath the leaves and these can easily be missed.

We have now reached the stage where to eat green beans with relish I have to do something more inventive than merely steaming them. It is surprising how few cookery books, even vegetarian ones, include a comprehensive section on imaginative ways of serving green beans. Look in the index under *beans* and you will find that the recipes invariably call for dried haricot or butter beans, or even tinned baked beans. Perhaps this is due to the fact that cookery book writers assume that you buy all your vegetables and, even in country areas where they are often sold at the gates of private homes, French and runner beans are quite expensive – unless you can live next door to someone like us, glad to give away almost as many as we consume!

Green beans make very good soup, both mixed with other vegetables and on their own, and can be added to casseroles and similar dishes involving chicken and turkey, which are cooked for shorter times than those containing red meat and therefore do not spoil the texture and flavour of the beans. Other than that, it is difficult to find ways of adding excitement to French and runner beans as a side vegetable. Runner beans can be enlivened if topped with cream flavoured with garlic and a sprinkling of chopped, very crispy bacon, and I sometimes cook runner or flat-podded French beans (many filet-type varieties are too tender and tend to go mushy if treated this way) under a lean lamb joint for the last half-hour before serving, to add a different flavour.

One popular way with us is to slice the beans into thin strips and cook until tender, then add 2oz (50g) chopped, fried mushrooms per 1lb (approx 500g) and mix well with $\frac{1}{4}$pt (150ml) double cream. The mixture is then seasoned with salt and pepper to taste and reheated gently without boiling, and is quite delicious.

Another is to boil thinly sliced green beans until cooked but still slightly crisp, then stir-fry them for a few minutes with four rashers lean, chopped bacon, a finely chopped medium onion, a large clove of garlic and two tablespoons light soy sauce per 1lb (approx 500g) beans. Apart from these particular family favourites, I am still experimenting to find something I would be happy to serve to Egon Ronay!

September

Tuesday 2

September is a pleasant month in the garden, a time for enjoying the products of all the hard work that has been expended on the land over the spring and summer, before the serious tidying and replanning of the coming weeks. Flowers are still blooming, but do not seem to need dead-heading nearly as much, and the leaves have yet to litter the ground. September is my time for catching up on tardy garden plans, magazine material which should have been submitted ages ago and, time permitting, seizing a moment away from home for recharging the batteries.

This morning I picked the first sweetcorn cobs. *F1 Honey Bantam Bicolour* is an early, short growing variety which, in ideal conditions, should be ready in August. Because my row is in partial shade during the afternoon it is about a fortnight behind in maturing. These more compact forms are becoming increasingly popular for amateur growing owing to their sturdy habit. Sweetcorn plants, especially the taller varieties, produce roots from above ground which help to feed and stabilize them if the soil is earthed up around the stems once or twice during the growing season. Side shoots appear from here which should not be removed, as they also help to nourish and steady the plants. Most shorter types thrive and remain stable without earthing-up and are therefore ideal for gardeners with limited time to expend on the maintenance of their crops.

Sweetcorn has come along a lot since it first started to become popular as an amateur crop. Older types took a long time to grow and mature. As the plants could not be planted outside until after the risk of frosts, it meant that only the most southerly areas of the United Kingdom stood a chance of producing ripe cobs. However, modern breeding programmes have seen the introduction of a wide choice of early-maturing varieties suitable for cultivation in all but the most inclement parts of the country, transforming sweetcorn from a luxury crop to one readily available to all.

When I first started growing sweetcorn, several years ago now, I made the mistake of having far too many plants, with the result that the freezer was jammed for ages with the cobs. Eventually I boiled them up, scraped the kernels off the cobs, and refroze them in manageable amounts for using in soups and prepared dishes, so releasing a lot of precious space. Now I only grow enough for immediate use, as there is something so special about juicy, fresh cobs in season, running in butter and drenched in black pepper, that I would rather wait twelve months for the experience than make do with frozen ones which, however well you do the job, never seem to be quite as good.

Honey Bantam Bicolour is one of the newer generation of sugar-enhanced sweetcorn, which have many times the level of sugar of normal varieties and therefore are deliciously sweet, and can even be eaten without cooking in certain circumstances. This variety is especially interesting in that it has kernels with mixed colours of cream and gold, so the cobs look particularly attractive when served.

Sweetcorn is an interesting enough plant to grow in the ornamental garden if there is enough space to plant a block. Block planting is necessary to ensure that the female flowers are adequately pollinated. The male flowers, or tassels, are produced at the top of the plant, the female flowers, or 'silks', are lower down, on the top of immature cobs. The female flowers are wind-pollinated, which is why it is more successful if the plants are grown in blocks where the pollen is able to blow from one plant to the others around it. Each plant produces at least one full-sized cob and many bear a second, smaller one which may or may not ripen depending upon the season. Older varieties may have more than one secondary cob, but it is unlikely that many of these will develop enough to be eaten. As it is usually possible to raise well over 20 plants from a packet of seeds, even if each only produces one top-quality cob, it is usually quite enough for the average small family.

To check for ripeness, the sheath surrounding the cob must be opened gently to expose the kernels. Ripe kernels exude a milky juice when scratched with a fingernail; the juice of unripe ones is clear and watery, and if the cobs have remained on the plant too long there is no juice at all. The kernels of very over-ripe cobs look dimpled and dehydrated – I have actually seen sweetcorn like this sold in a local supermarket when it should have been rejected weeks before. It is essential that the cobs are checked regularly, as over-ripe ones are dry, doughy and unappetizing when cooked.

The kernels I tested this morning were just right, so I picked about a third of the crop – eight in all, which should last us until the end of the week. The good thing about these super-sweet forms is that the sweetness and flavour does not deteriorate for several days after picking, unlike conventional varieties which rapidly lose sweetness and flavour once detached from the plants. The first cobs in the row had, disappointingly, not been completely pollinated, so some of the kernels had not developed. As it is not worth serving these as a starter, I scrapcd the kernels off the cobs and I shall serve these, raw, with a nice salad of *Little Gem* lettuce, *Sungold* tomatoes and *Radar* French beans in vinaigrette, topped with chopped *Santa Clause* onions – the red-tinted stems are great for visual effect. Delicious!

I think it unlikely that many of the secondary kernels will come to anything. In fact, most of them have only grown to the size of the 'baby sweetcorn' so popular for stir-fries and dressed salads. Out of interest, I picked one and tried it to see what it was like to eat. In fact, it was surprisingly like the baby corn which costs so much to buy, tender and tasty, so these undeveloped cobs will not be wasted,

Lettuce – red and green

116

and over the next week or so I shall add them to *Sugar Snap* peas and finely diced carrots for interesting mixed vegetable servings.

Friday 5

Today involved a trip to the area behind the Lincolnshire coast to inspect a garden which has been an on-going project of mine for several years – the children were small when I started, now most of them have flown the nest! The landscaping has been done a little at a time, when time and finances allowed, during which period my client has added more land, so it is one of those ventures which seems never-ending. The main object of the visit today was to inspect a field which was planned as a wild-life area, with native trees and shrubs, a pond and wildflowers. My involvement was only to supply the plan. Because my client was excessively busy at the time, he called in a local 'landscape gardener' to supply the plants and provide the labour. As has so often happened to me in the past, the wrong trees and shrubs were supplied, and then planted in all the wrong places. Today I had to decide which species could be left and which had to be replaced, which could remain in the positions in which they were erroneously planted and which would have to be removed once dormant to preserve the effect of the design. Wherever possible, I tried to save time and money by compromising, to be too pedantic at this time would have caused incalculable inconvenience. It is very annoying when this type of thing happens, as my client will have spent twice as much as was necessary by the time he has put things right, owing to the inefficiency of the contractor.

Visiting this garden always reminds me how grateful I should be that I live in the fertile siltlands in the south of the county. Like much of the land which lies behind the coast in that area, the soil is heavy clay, poorly drained in winter and like a brickfield in summer, with biting easterly winds in late winter and spring – don't I know it, as my previous home was in that area! Consequently, plants grow at less than half the rate that they would in more favourable conditions and the mortality rate is high. We generally try one replacement with the same species each time something dies, then we abandon that idea and try something tougher. Establishing this particular garden, therefore, is very much a matter of trial and error.

After we had sorted out the field and various other problem spots, my client proudly showed me his vegetable garden. He will not use chemicals under any circumstances, so the caterpillars had had a field day on his brassicas, most of which had been completely skeletonized,

Dwarf beans make a useful stop-gap between broad beans and runners

much as mine would have been if I had not sprayed with permethrin last month. The type of planting we have used around his house and in the field is a haven for butterflies of all types. My client is convinced that now the cabbage white caterpillars have pupated, the sprouts and broccoli will recover. I did not want to hurt his feelings, but I feel he is in for a great disappointment. To produce first-rate crops from this group of vegetables, it is essential that they grow without a check – removing all the leaves except for the mid-ribs and main veins can certainly be considered a set-back, to put it mildly!

There was time after we had finished for a picnic by the sea. Carl loves the empty, flat beaches of this part of the country and gallops for miles in true greyhound fashion. After earlier trials and tribulations, he has settled down to make a very nice pet – quiet and gentle, always grateful for small mercies, but not to be trusted with anything edible, as he is a terrible thief.

Saturday 6

It is difficult to keep up with the tomatoes as they ripen so I am now freezing most of them. I freeze them raw, merely washed and dried before packing into polythene bags. *Sungold* is a very watery tomato and is not particularly suitable for using in cooked dishes other than soups, for which it is ideal. I have a very good soup recipe even for people who, like myself, do not especially like tomato soup. I think that my problem is that as a young child I had, of necessity, to live with my aunt for several months. Childless herself, she made it quite obvious she did not enjoy looking after her sister's offspring and put herself out as little as possible when suddenly confronted with a

The Sungold *tomato is ideal for soups*

youngster. As far as I remember, apart from the vegetables my uncle grew in the hallowed patch at the bottom of the garden, I seemed to live on a diet totally consisting of tinned baked beans and canned cream of tomato soup, which was only relieved by the Saturday treat of fish and chips from the local chip shop. It is hardly surprising, therefore, that to this day I can happily give these foods a miss, although I can occasionally fancy a small portion of fish and chips if I am very hungry.

My tomato soup recipe consists of frying together 1 thinly sliced small onion or 2 large shallots, 1 clove of garlic and 2 finely chopped sticks of celery, with leaves if possible, for around 7–8 minutes until softened but not browned. I then add about 2lb (1 kilo) frozen *Sungold* tomatoes, a bay leaf, 2 cloves, $\frac{1}{4}$ teaspoon basil (dried will do), 1 tablespoon coarsely chopped fresh parsley and 2 teaspoons sugar, and simmer it all together very gently for about quarter of an hour until the vegetables are thoroughly cooked and soft. This is then rubbed through a sieve to extract all the liquid and pulp and remove the fibrous material and tomato skins. If necessary, it is then made up to 1pt (600ml) with water – the liquid content of the vegetables, especially the tomatoes, tends to vary. Salt and freshly ground black pepper are added to taste and then the soup is reheated with the addition of 2 teaspoons of lemon juice. It can be served hot or well chilled with a swirl of cream in each bowl if liked, and sprinkled with finely chopped parsley. This will serve four as a starter, or two as a snack meal, and is delicious with hot garlic bread. Our local organic flour mill, the Maud Foster Mill in Boston, sells onion flour, which is strong white flour with dried onion flakes, and this makes a scrumptious savoury bread which is a very good alternative to garlic bread with soups such as this clear tomato one.

Wednesday 10

The *F1 Caravel* calabrese plants have some good, solid heads on them. I removed all the main heads – five of them – to encourage smaller side shoots to develop, and froze them. Calabrese freezes extremely well, in fact better than cauliflower; if blanched for the minimum length of time (see Appendix 1, page 153) the head remains nice and solid. The adjacent Brussels sprouts have small sprouts on the lower parts of the stems which could almost be cooked as the 'baby sprouts' which are sold in supermarkets at such exorbitant prices. However, as other vegetables are in such good supply, I shall allow these to develop into full-sized ones.

Caravel *broccoli*

I drew the earth up round the stems of the *Walcheren Winter Pilgrim* cauliflowers to ensure that they are as firmly established in the ground as possible before the winter winds. The greatest cause of cauliflower failure – small, open curds with a poor texture and flavour – is loose planting. Plants which flap about in the wind are unlikely to do well, and this applies to Brussels sprouts, too. While in the mood for earthing-up, I also gently pulled a little soil round the

121

base of the leeks. *Winora* usually produces a good length of white base without too much earthing up, especially if properly planted in individual holes in the bottom of a trench which gradually fills in during cultivation between the rows. Too much ground disturbance, like earthing-up, can cause soil to trickle down between the leaves, making cleaning difficult to well nigh impossible at cooking time, so this is the one and only time this particular crop will be earthed-up before harvesting.

Monday 15

We have just returned from hosting a garden lovers' weekend. I have been working for this small Lincoln-based company for many years in this capacity, and thoroughly enjoy the work, which is so different from every other aspect of my life. The job consists of accompanying a party of like-minded individuals on short breaks, based on a good hotel which is convenient for visiting a number of gardens, some well-known, some private, over the course of a two- or three-day weekend. My main function is just to be there, to answer questions, chat about my work in general and, hopefully, supply information on the gardens we are seeing. The first evening, after what is invariably an excellent meal, I also give our party a short lecture on whatever subject I think is likely to be of interest to them. Because many have travelled long distances and are now full of food, drink and good cheer, the talk is of necessity light-hearted or, as I always jokingly tell them, one they can snooze to, providing they do it quietly, as loud snoring is liable to upset my train of thought!

The following evening, because by this time the group has relaxed and settled down, there is another lecture of a slightly more in-depth nature. The days between are filled to overflowing with beautiful gardens linked with interesting coach journeys, and yet more good food, so, as one can imagine, these short holidays are very popular.

This time, our weekend was based in Derby, and covered gardens in parts of Leicestershire, Nottinghamshire and Derbyshire which do not immediately spring to mind as essential holiday locations. Because of this, it was unknown territory for many of our guests, and allowed me to see from a coach many places which, until then, I had merely sped through in the car, en route to somewhere else. It is a part of the country with which we are both reasonably familiar – John was born near Derby – so in addition to all the gardening information I was called on to impart, we were able to supply some local colour as well.

The surprising thing about many gardens which are available to the public is how few of them have interesting kitchen gardens, unless they are National Trust or English Heritage properties where these are a vital part of the original landscaping and have been, or are in the process of being, restored. Highlight of this weekend in the vegetable gardening category was the kitchen garden of Calke Abbey near Ashby-de-la-Zouch in Leicestershire. This is a house with a time-warp, where the vegetable gardens have been extensively renovated since coming under the National Trust, and which have improved considerably since my first visit about three years ago. It was fortunate that we had eaten first, as the mouth-watering crops would have been torture to behold otherwise. I am sure that if more gardens open to the public had inspiring kitchen gardens, it would encourage more people to try this highly rewarding, but still sadly less than popular, aspect of gardening for themselves.

Thursday 18

True to form, we hardly seem to be at home at all at the moment. Yesterday it was London and the *Jimmy Young Show*, and another opportunity to reduce my surpluses of tomatoes and runner beans. My climbing beans are at last beginning to slow up, and the French beans are just about finished, but the tomatoes, if anything, are more prolific than ever. Today John is back at Radio Lincolnshire, also heavily laden with bags of produce. On Saturday I leave for the gardening trade show in Birmingham, so today has been spent in getting ahead of myself on the vegetable patch so it does not get the better of me in my absence. Tomorrow the cottage garden at home will get the same treatment.

We ate the last of the sweetcorn earlier this week, so I was able to remove the old plants. The easiest way to dispose of these is to cut them into short pieces and add them to the compost heap, then cover them up with the latest supply of yellowing old leaves from the brassica plants which are wet and slimy and therefore mix well with the fibrous stems of the sweetcorn.

The tops of what is left of the *Nadine* potatoes have died down and the skins have set, so there is little point in leaving them in the ground any longer. Just under half of the crop is still left so I lifted them and washed them with a jet from the hose before leaving them to dry in the sun for a few hours. They were then ready to put in a hessian sack which will be kept in John's workshop in the shed, which has a constant temperature of around 50°F (10°C). Potato tubers left in the light for more than a few hours will start to turn green. The green material contains substances which, although they will not kill you unless you consume large quantities, are likely to give you an upset stomach if you eat more than a very small amount. It is essential, therefore, to get them into the dark as soon as possible after lifting, but they should be cleaned first to remove traces of any diseases, and thoroughly dried to prevent rotting. Any damaged ones should not be stored but used immediately – one or two of my tubers had slug and wireworm damage, and I managed to spear several of the best by careless lifting – but the rest will last us through the next few weeks, as we do not normally eat large amounts of potatoes. This meant that the area between the west shelter belt and the beetroot was completely empty, so I rough-dug it in preparation for winter weather action.

We take the caravan to a small site near Coleshill for the exhibition which, as it has grown over the years, has become a 'must' for everybody involved in the gardening and associated leisure industries. Betty, the lady who looks after the site, is a dedicated gardener, especially where vegetables are concerned, and she and her husband regularly produce larger surpluses than I do. Unless I have work in the Birmingham area, we only see each other once a year, and so there was plenty of news to swap and tips to exchange. It is only a short drive down the road to the National Exhibition Centre. There is so much to see that we do not feel like wining and dining out after the long days trying to view every stand, so the comfort of our own home-on-wheels is particularly important. Much of this afternoon was taken up with loading provisions to last five days – beans, beetroot, potatoes, carrots, tomatoes, lettuce, courgettes, onions and broccoli should provide us with a balanced diet and supply me with enough ingredients to indulge my culinary extravagances should I be so inclined.

Wednesday 24

The show was bigger than ever – trade only, so visitors were able to move about freely and investigate everything there was on display, unlike at many public shows, especially Chelsea Flower Show, where the guests are crammed in like sardines and it is impossible to see anything unless you are up with the lark. It is not a show where I am likely to buy a lot as it is aimed mainly at retailers, although I frequently end up with a new mower on the last day when many of the display goods are sold off cut-price, and this year was no exception as the small mower I use for home is now on its last legs. This means that our mower count to date is six, though not all are in working order! My main interest in this type of exhibition is seeing what my clients, readers and audience are going to be offered for sale in the coming year, to inspect the products first-hand, and sort the wheat from the chaff when it comes to quality, usefulness and value for money. I arrive home buried under a pile of leaflets which I always intend to read, mark, learn and inwardly digest but which generally lie unopened until the next show, apart from those relating to items of importance needing further investigation as the new year unfolds. This leads me to think that the amateur gardening public is offered many articles it neither wants nor, in fact, really needs, and so part of my job must be to offer guidance in this respect and undertake the research involved. I now have two very sore feet and aching calves to show for it. Such are the occupational hazards of the conscientious gardening adviser.

Saturday 27

My 'ex' has decided he wants coving in his living room and bed-rooms, and John has agreed to do it for him over this weekend. John will do the skilled work, Tony will labour and, inevitably, I shall cook and make sure everybody's stomach is full. Tony is a great fan of run-ner beans and was most impressed with the ones I arrived with this morning. He showed me the remains of some he had bought in Stamford earlier this week and once again I realized how lucky I am to have a more or less unlimited supply of such vegetables. He was even more impressed with the last of the *Mokum* carrots I took along as well, as he had forgotten what non-commercially grown carrots could taste like. My quick tomato soup will come in useful later in the project, as I know I shall end up doing the painting – I am much neater, quicker and more professional than the two chaps, so my time for cooking will be severely curtailed. Ah, what it is to be multi-skilled!

October

Thursday 2

Our new next-door neighbour recently hired a 'tree surgeon' to thin and generally sort out a large number of overgrown trees on her plot. The result was a disaster – many lovely and useful trees removed altogether and the rest given a chain-saw haircut from which I fear they will never satisfactorily recover. A further outcome was that the fence between us, which in fact belongs to us, has been loosened – it is our fence and it may indeed be a blessing in disguise, as I knew the posts were beginning to rot at the base, but we really did not want to start doing major boundary works at this moment. However, as it is covered with climbing plants and we could be subjected to strong autumn winds at any minute, we have no option but to undertake what is quite a big repair.

The previous owner (or his contractor) made the costly but common mistake of sinking the wooden fence posts in concrete; although this makes a firm job at the time, premature rotting occurs at the point where the post enters the concrete. The easiest, most efficient way of repairing a fence affected in this way with the minimum amount of disturbance is to bolt the bottom of the post above the soil to a short concrete spur which is itself buried in concrete below ground level. As today was fine, warm and still, it seemed a good time to make a start on the first of the three broken posts.

John and I can work together quite happily on most jobs, but I knew from the outset that this was not going to be one of them. My overcrowded vegetation brings out the worst in him, while I wince every time his size 11 shoe makes contact with one of my cherished possessions. I felt it prudent to take myself elsewhere, so I took a trip to the vegetable patch and cut a large bag of parsley before it becomes damaged by the winter weather. I washed and froze most of it as I find this is the easiest way of using it in cooking. Fresh parsley will keep up to a week in the salad drawer of the refrigerator; it makes a delicious accompaniment to many kinds of dishes, particularly fish, if it is fried in very hot oil for a short time. If the oil is hot enough, the parsley becomes crisp, rather like the 'seaweed' of Chinese cookery (which is not seaweed as such at all, but a kind of broccoli), and can be sprinkled over the dish in question immediately before serving.

Saturday 4

It is hard to believe it is October, as daytime temperatures are still in the sixties. It was too good a day to miss, so we spent the day on the Patch, digging over the areas vacated by harvested vegetables. It is time to plant winter-hardy Japanese onion sets, and as we are hoping to have a few days' sailing on the Norfolk Broads next week, I felt I ought to get on with the job, so I popped down to the local garden centre and bought a pound of sets. Winter-hardy onion sets are planted in exactly the same way as those planted in spring for a main crop. In our plot they now occupy the area where the broad beans grew, as the first crop of the second year of the crop rotation. The nitrogen fixed in the soil by the roots of the beans will give an early boost to growth for the onion bulbs. It was quite warm enough to eat outdoors – I took a couple of lamb chops out of the freezer before we left the cottage and these went well with baked *Nadine* potatoes, a serving of runner beans, which are still cropping well, and a modest portion of secondary shoots from the *Caravel* broccoli. I cannot help but feel that this weather is unlikely to last much longer, but a short break under sail will be a real treat if it does.

Monday 6

Packing to go away for a week's sailing is always quite an undertaking. As well as enough clothes to see us through the week, including warm woollies and plenty of suitable changes in case we get wet (we haven't yet, but there is always a first time!), I like to take sufficent food to ensure we do not need to shop or eat out if we do not want to. First thing, therefore – another lovely, warm, sunny morning – I called in at the Patch to collect all the vegetables I thought we might need. I stripped the climbing beans and French beans of the remainder of their pods, even small ones in case there is a frost in our absence, cut the first *Bingo* cabbage, which was a beautiful, solid specimen and dug up around 2lbs (1 kilo) of *Bertan* carrots. With these and a carrier bag full of potatoes, onions and shallots, some young *Little Gem* lettuces, two largish *Greyzini* courgettes, a nice bunch of *White Lisbon* onions and a large bag of *Sungold* tomatoes, we should surely have as much as we need. There is a small amount of blight showing on one or two tomato plants, so this might be the last we get this season.

Tuesday 14

My premonition about the weather proved to be entirely correct. No sooner had I stocked up with food for the holiday than the sky clouded over; by 4pm it was raining heavily, and it was so wet and windy the following day that we had serious doubts as to whether we should bother. It is not just a question of hopping into the car and disappearing. John takes the caravan and fourtrack, and I follow with the car towing the Wayfarer dinghy and all the sailing gear. For such effort, therefore, we have to be quite sure we are going to spend a decent amount of time on the water.

By lunchtime we decided to risk it, and although the drive was filthy, the weather had improved sufficiently to allow us to launch the boat, minus sails, and potter up the river for a short while under the outboard motor. This was the first time Carl had been on the water; he was totally terrified, and spent most of the trip sitting on my knee!

This, and a journey under sail last Saturday which ended in our becoming totally becalmed, was the only chance we had to enjoy the Broads this trip – every other day was too wet, too windy or too still. It was lucky that I had stocked up well with provisions, as the holiday turned out to be more of an opportunity for culinary experimentation than anything else.

Bingo is an ideal cabbage for coleslaw. My coleslaw recipe may not conform to the general idea of this dish, but it certainly goes well with most salads and cold meats and dishes. I shred about ½lb (225g) cabbage finely, and combine it with a finely grated large shallot, 2 grated medium carrots (the average *Bertan* is what I call a medium carrot), a chopped apple sprinkled with lemon juice to prevent it from turning brown, a small handful of fresh, chopped parsley, and salt and freshly ground pepper. This is mixed with enough mayonnaise (I confess I use ready-made for this) to make the mixture spoonable but not sloppy. Even the anti-cabbage John will eat cabbage prepared in this way, providing the portions are not too large.

The drawback I have found with *Sungold* tomatoes – splitting skins – has become much worse as the season has progressed. By the end of our week away, we were left with a lot of perfectly edible fruit, but which could not be served in salads because they were becoming sloppy. It seemed a shame to throw them away as they may well have been the last of the year, so I boiled them up and then passed them through a sieve. This gave me nearly 2 pints (1 litre) of tomato juice which, chilled, was very tasty with a measure (or two) of vodka and a dash of Worcestershire sauce. After many wet, windy hikes along the beach experiencing the turbulent North Sea at its very worst while we waited in vain for the sailing opportunities which never came, my home-grown Bloody Marys were most acceptable!

We had to return home this morning as I have a charity lecture tomorrow evening. The weather shows no sign of improving, and it was a treat to get back to our log fire.

Saturday 18

In our absence, tomato blight had spread through the whole row of *Sungold*. The plants were limp and blackened, and the huge crop of unripe tomatoes hung in long, disgusting skeins or littered the ground. To prevent infection carrying over from year to year, it is important to clean up as much affected material as possible. I pulled up the plants, raked up as many of the fruit as possible, and bagged up the lot. Material of this sort should never be composted, and the soggy condition of the vegetable garden and all its rubbish made burning a non-starter, so I reluctantly took the hapless tomatoes to our local recycling plant.

Years behind most other local authorities, our council has suddenly realized that green waste should be treated differently from other household waste, so it is segregated and taken away for shredding and composting. In this case it would have been anti-social of me to have dumped my blight-ridden haulms in the green waste container, as this could help to over-winter the disease on somebody else's doorstep, so the tomatoes ended up in the skip for waste which is totally unrecyclable in any shape or form. It is always a dilemma knowing how to deal with infected plant material, but of many ways round the problem, this seems the least unsatisfactory.

Tuesday 21

As it was a reasonably dry morning, I thought it was time to lift the rest of the beetroot *Mondella*. Some years I have left the beetroot crop in all winter, but there is not all that much left, so I thought I would make a beetroot salad to go with some red lettuce leaves and the excess could be cooked and pickled when I had a spare moment. A very easy beetroot salad can be made by dicing around 1lb (454g) of cooked beetroot and combining it with ½pint (300ml) yoghurt or soured cream, flavoured with a crushed garlic clove and freshly ground black pepper. Serving on a bed of red salad leaves makes this pink creation look particularly attractive.

Thursday 22

I always think it is a pity that so many people detest Brussels sprouts. Maybe it is because they are usually cooked far too long, or perhaps it is because they are allowed to opt-out of the sprout servings when they are youngsters, but those who refuse to eat sprouts really do not know what they are missing. For me the best sprouts of the year are those of the first picking. Unless we have a glut, I do not usually freeze them, which means we never eat sprouts out of season, so they come as a welcome change at this time of year. This season the *Icarus* crop is undoubtedly the best I have ever grown – medium-tall stalks packed with lovely firm sprouts. I could have started picking a month ago, but I wanted the whole crop to reach a good size, the average now being just over an inch (2.5cm) across, which I think is about the optimum diameter of a decent sprout. There are hardly any loose, 'blown' sprouts at the base of the plants, so virtually every one is harvestable.

Possibly the easiest and most satisfactory way of harvesting sprouts is to dig up the whole plant, wash the roots and hang it,

upside down, outside by the kitchen door. The sprouts remain fresh, green and firm until they are needed, and picking is so much more pleasant.

It is often said that sprouts taste sweeter and generally have a better flavour once they have had a frost on them, but with these modern, super-sweet, Vitamin C-rich varieties such as *Icarus*, I do not find this applies any longer.

There are many ways of serving Brussels sprouts to make them more exciting, but with the first pickings of the year, there is nothing, to my mind, more appetizing than steaming them for around five minutes, so they are just cooked but still crisp, and then serving them with plenty of freshly milled pepper, a good knob of butter, and a little ground sea salt. Yummy!

Friday 24

I had to visit my solicitor friend, Dianne, this morning. Her partner's son had recently completed his projects for his Duke of Edinburgh Gold Medal Award, one of which was to create a garden, including fruit and vegetables, apparently from a wilderness – as he had been allocated a patch behind the office which, frankly, I would not have wanted as a first excursion into the pleasures of gardening. He had done sterling work in the circumstances, so I willingly signed his book, but I am still curious why, knowing Dianne to be vegetarian, he had not thought to offer her some of the fruits of his labours to sample. Perhaps he thought she grew her own, but Dianne is much too busy unravelling the complicated lives of her clients to find time to grow her own vegetables. This is why I find her so useful, as I can offload all my surpluses onto her. The runner beans are still cropping prolifically, so I took along about 1lb (nearly half a kilo) of these, together with a *Bingo* cabbage weighing nearly 3lb (around 1½ kilo), some shallots, a stalk of *Icarus* Brussels sprouts, and some magnificent red stems of rhubarb chard.

Saturday 25

Ruth and David, our friends who relieve me of my surplus vegetable plants, came for supper this evening, and suggested that as they are on their half-term holiday, we should take a short break together in our respective caravans. As I am working for Granada Television next Thursday, and we have a horrendously early start, it seemed sensible to choose a venue within easy driving distance of Manchester, and decided on the Derbyshire Dales. I really cannot afford any more time away, but the opportunity of a few days in Derbyshire in the autumn was too tempting to turn down. We have decided to leave on Tuesday, so the next two days will be busy ones.

Sunday 26 and Monday 27

However much ready-prepared food I buy during the year, Christmas is a time when I like to make everything myself, right down to the turkey accompaniments and brandy butter. Most of our visitors want pickles with everything (apart from the Christmas roast, that is) so it is high time I got down to stocking up the store cupboard. Pickles and chutneys always benefit from being allowed to mature for a few weeks and leaving the job until I return from Derbyshire will not give enough time for this before the holiday season starts. Apart from pickled onions, which are essential, my choice of pickles depends on what is in plentiful supply at the time – this year it is the remainder of the beetroot and, for a change, I thought I would make a few jars of a tasty and unusual cabbage pickle which very much resembles piccalilli in flavour, but not in texture. As most pickled vegetables require treating with salt for 24 hours to remove excess moisture, one needs two consecutive days for a pickling session.

Pickled onions

Ingredients

2 quarts pickling onions or shallots

1 quart white pickling vinegar

1 tablespoon mixed pickling spice, plus a few extra black pepper-corns and cloves

$\frac{1}{2}$ cupful salt

$\frac{1}{2}$ cupful sugar

Method
Peel the onions or shallots and place in a basin, mixed with the salt, overnight. Rinse to remove all salt, and dry. Boil the pickling vinegar, sugar and spices together for a few minutes, add the onions, and bring back to the boil. Pack the onions tightly while still hot in clean, warm jars and fill up to the brim with the spiced vinegar – do not remove the spices as these look attractive mixed amongst the onions. Cover when cold. A mixture of golden *Atlantic*

and red *Pikant* shallots is very decorative, especially when white vinegar is used, although the flavour is just as good with brown malt vinegar. If ready spiced vinegar is not handy, ordinary white or brown malt vinegar can be used, but you need about three tablespoonsful of pickling spice to give a really good, spicy flavour. These pickled onions need at least two months to mature before eating.

Quick pickled beetroot

Boil the beetroot until soft, then remove skins, slice thinly and pack into jars. Fill up with hot, ready-spiced pickling vinegar, leave to become quite cold, then cover. Pickled beetroot has a very short shelf life before it turns mouldy, so it should be kept as cool as possible and be consumed quickly. I find that using hot vinegar rather than cold, straight from the bottle, does prolong the life considerably.

Green cabbage pickle

Ingredients

1 large cabbage (I found a whole head of *Bingo* just the right quantity)

4 large onions

$\frac{1}{3}$ cupful salt

1 quart white spiced pickling vinegar

1 cupful plain flour

2 cupfuls granulated sugar

3 teaspoons mild Madras curry powder

2 tablespoons mustard powder

1 pint white malt vinegar

Method

Shred the cabbage and onions, finely sprinkle with salt, and leave for 24 hours. Drain and rinse well through a sieve, then empty onto a clean, dry teatowel and dab to remove excess water. You will find that the cabbage still tastes very salty, but this over-saltiness disappears when the pickle is cooked. Simmer slowly in the spiced vinegar for 15 minutes. Combine the flour, sugar, curry powder, mustard and vinegar, then pour over the cabbage and spiced vinegar and boil together for about five minutes or until the mixture thickens. Fill clean, warm jars with the pickle and cover when cold.

I use the metal or plastic screw-tops which come with jars of bought jam, peanut butter, marmalade, etc, as this prevents the vinegar evaporating after a while, so making the pickle dry.

Ruth and I take it in turns to provide the evening meal when we are away, so it was necessary to check the Patch to see what fresh vegetables were available. Apart from another good yield of runner and *Romano* climbing French beans, it seemed that courgettes would be the main vegetable on the forthcoming menus, plus a *Greyzini* courgette which had earlier escaped my attention and turned into a marrow. I added this to the collection in case we ran out of fresh vegetables. If all else failed, we could make a savoury stuffing and eat it as a main course, but I suspect it might not be the first choice, especially of the men.

October 31

After four days of pleasant autumn weather, winter arrived last night, with clear skies and a penetrating frost. Our site is in the middle of a young wood, and all around us was the patter of leaves as they let go of their hold of the rime-spangled branches. After a day in the Dales, a substantial meal was essential.

The courgettes needed a disguise before John was likely to relish them. One appetizing quick way is to fry a sliced onion and a finely chopped garlic clove in a knob of butter until soft, then stir-fry 1lb (454g) sliced courgettes until hot but still crisp. Add a teaspoon of Italian herbs, salt and pepper to taste, and enough tomato ketchup to coat the vegetables thoroughly, and the result is an easily prepared, savoury side dish.

November

Monday 3

The low temperatures last week were obviously not confined to the Derbyshire Dales as they have certainly had their effect on the vegetable garden in our absence. All the bean plants are in a sorry state, and what is left of the parsley is now lying in a flattened, soggy mess. I am now faced with disentangling the plants from the supports, but it will have to be much pleasanter weather than the raw, foggy day we have today before I tackle that fiddly, time-consuming and finger-freezing job.

Now we have had a frost, it is time to start lifting the *Gladiator* parsnips, which always taste better after a period of cold weather. I generally undertake the first lifting with some trepidation, as it is impossible to tell how successful the crop has been until a sample is dug – last year was a disaster, with very few roots thicker than my finger. I put this down to the fact that parsnips need a very long growing season and, because of an unusually late spring, it was impossible to sow them until the beginning of May. Then we had an exceptionally dry year, and it was difficult to irrigate adequately.

However, this year I need not have worried, as the roots I lifted were acceptably thick and long. They would have been even thicker if I had thinned the row, but I prefer to disturb the seedlings as little as possible, and the crop is quite adequate for normal purposes.

I have to confess that parsnips are not my favourite vegetable and, left to myself, I do not think I would bother to grow them, but John loves them, and so do most of our friends, so I feel obliged to have a token row, which usually lasts us well into the spring. I usually par-boil them for about three minutes and then roast them with potatoes around a joint of meat, although for a change I sometimes mash boiled parsnips with carrots, a large knob of butter and a seasoning of salt, pepper and a pinch of nutmeg.

Surprisingly, the lettuces and 'spring' – or to be more accurate at this time of year, salad – onions have survived the frost remarkably well. My vegetarian solicitor friend Dianne and husband Bill are coming to dinner on Saturday evening. I shall take the easy way out and have pizzas and quiches with a green salad, some kind of potato concoction and, to start with, a vegetable-only soup.

Wednesday 5

There are some lovely heads on the *Romanesco* broccoli. It is possi-bly my imagination, but I am convinced that the attractive, pale green, pointy heads of these particular types of broccoli taste com-pletely different from every other kind. *Romanesco* caused a lot of comment when it was first introduced; now it is an old favourite, and there are several other similar varieties, yet you still only see it for sale in the specialist greengrocer and better supermarket. There is no better way of cooking than lightly steaming or microwaving it, then serving with a knob of butter, ground sea salt and freshly milled black pepper.

Sunday 9

We had a most enjoyable evening yesterday. We had not seen Bill and Dianne socially for a very long time, and there was much catching up on news and gossip. The soup I eventually opted for was carrot and mint, which is not too filling and very easy to make, and there were no garlic potatoes left at the end of the meal. Dianne is having a party in January for a Big Birthday – I shall be interested to see how she manages to fit the number of guests she is inviting into her cottage. Admittedly, it is twice the size of ours, but more than four extra people in our cottage and we begin to feel like sardines in a can.

Carrot and mint soup

Fry together 1½ lb (675g) thinly sliced carrots, a finely chopped medium onion and a chopped clove of garlic for about 5 minutes in 1oz (25g) butter, or a tablespoon vegetable oil. Add 1 pint (600ml) vegetable stock made with 2 stock cubes and 1 pint (600ml) milk, and simmer for 15 minutes. Liquidize until quite smooth, then add 1 teaspoon Worcestershire sauce, salt and pepper to taste and 2 tablespoons finely chopped spearmint or apple mint. Serve with a swirl of cream or natural yoghurt. This also makes a good summer soup if served well chilled.

Garlic potatoes

Serves

Ingredients

1½ lb (675g) potatoes

1 large onion

3 cloves of garlic, finely chopped

Salt and freshly ground pepper

¼ pt (150ml) hot vegetable stock, made with one vegetable stock cube

2oz (50g) grated strong Cheddar cheese

Method

Slice the potatoes and onion and chop the garlic cloves finely. Place layers of potatoes and onions sprinkled with the chopped garlic and plenty of salt and freshly ground black pepper in a large, greased, ovenproof dish. Pour over the stock, cover and cook in a preheated oven at 190°C (375°F, gas mark 5), for 40 minutes. Uncover, sprinkle with the Cheddar cheese, and cook for a further 30–40 minutes, until golden brown and bubbling.

Tuesday 11

The frost at the end of last month has intensified the colour in the stems of the rhubarb chard to a gor`geous ruby shade. While the leaves themselves, which have turned a deep purple-green, are rather too coarse to use, the mid-ribs will still make a tasty and eye-catching dish. Because the flavour is now quite strong, they are best

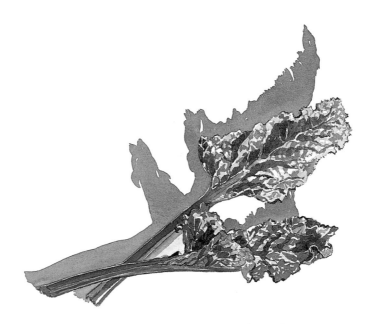

steamed until soft and served with a Hollandaise sauce. A quick Hollandaise sauce can be made by putting three egg yolks and a tablespoon of lemon juice with salt and freshly ground pepper into a food processor for a few seconds, then adding 4oz (114g) hot melted butter. The mixture is blended until thick and served immediately but, if liked, 3 tablespoons of lightly whipped double cream may be mixed in gently before serving.

Thursday 13

Today was fine and fairly warm, so I decided it was time to clear the dead climbing bean haulms and their supports. There is no easy way of disentangling the bean stems, other than cutting the plants off at the bottom and then cutting and picking off every single bit of twisted stem from every single pole. Like the pea haulms, these are a useful addition to the compost heap as they contain a lot of nitrogen and, as with the peas, it is better to dig in the roots, plus their nitrogen-fixing bacteria, to provide a boost to the crops in next year's rotation.

The plastic-coated metal poles were easily cleaned with a cloth wrung out in Jeyes' Fluid. The task was so much easier than in previous years, when I had to scrub all the canes before tying them in bundles to store.

There was very little left of the dwarf bean plants after the frost, so they could be dug in in their entirety. They should have disappeared by the time I come to making seed beds next spring.

Monday 17

The sprouts are so fine it is tempting to pick and cook them indiscriminately but, with only 15 plants, I shall have to ration myself if they are to last into the new year. One stem per week from now onwards is our limit – this provides about six modest portions.

Wednesday 19

I lifted some *Bertan* carrots this morning and found soil pest damage – mainly carrot fly. Usually I leave maincrop carrots in the ground all winter, but if there is any damage I find it is better to lift them and, rather than try storing the roots which will probably rot, I freeze them in usable amounts. After preparation and blanching, I froze about 6lb ($2\frac{1}{2}$ kilo), which, with those frozen earlier, should see us through the winter for soups, stews and other prepared dishes. If required as a vegetable accompaniment to a main dish, they are best cooked and mashed with salt, pepper and dill.

Friday 21

I gave a lecture to a local horticultural society this evening which, as usual, ended with a lively question and answer session. One questioner was concerned that when he pulled up his runner bean plants, the roots of many of them were swollen, rather like those of dahlias, and he wondered if they were suffering from a nasty disease, akin to the club root disease so often experienced with the cabbage family. He was both surprised and relieved when I assured him that runner beans are actually perennial, and do produce tuberous roots once mature enough. In fact, if you ever have a plant which is particularly special and has managed to produce tubers, it is worth lifting the root and storing it in just-damp sand over winter, for replanting after risk of frost next spring. Many years ago, when I had a small vegetable plot in our cottage garden, I managed to keep the same runner bean plants going for three years against a sunny fence, by covering the base of the plants with a thick mulch every winter. When I did eventually lift them, the tuberous roots were as large as those of well-grown dahlias.

Sunday 23

I had an advance Christmas present today – a new microwave oven. We have had a microwave oven for about eight years, and it does cook vegetables extremely well, but unless I refer to the instruction book each time, cooking is largely a matter of trial and error. There was really nothing wrong with the one I was using, but I felt that technology had moved on apace since I bought mine. I needed a cabinet colour to complement the kitchen, which has been revamped since my first one was bought, and the present one would be very useful in our kitchen at the Patch. I was amazed at what was available – everything cooks by sensor at the touch of a button, and I am sure this will revolutionize my vegetable cookery in the months to come.

Wednesday 26

There was cold meat for lunch today, and temptation overcame me to open a jar of pickled onions, even though they should have been left to mature for another month. John went out for the afternoon and returned looking grey and ill, with a stomach ache which, we speculated, was probably due to a virulent form of food poisoning. During the evening, it became apparent it was nothing more than a very bad attack of indigestion and wind. This must be a lesson to us: if pickled onions need two months to mature, then two months it has to be!

Friday 28

Although the weather is still uncommonly mild for this time of year, frost, snow and bitterly cold conditions are forecast. There were still some good *Romanesco* spears on the plants which could be damaged by very severe weather, so I harvested them and froze them for future use. As the ground was workable, I removed the old plants and rough-dug the space they had vacated. This means that in that area, only the *Rudolph* purple sprouting broccoli remains. This will be there for a good while longer, as I am not expecting to see the first spears before the middle of January.

Sunday 30

The marrow which started life as a *Greyzini* courgette is still sitting on the kitchen windowsill, which shows just how popular marrow is in our family. I must do something with it shortly, before it explodes. However, it has proved what an excellent marrow *Greyzini* will produce if left alone to get on with it.

December

– A Year On

Tuesday 2

I am amazed that despite all the wet, cold weather and frost we have had recently, the lettuces and salad onions are still surviving, and so we were able to have a home-grown green salad for tea. The *Tundra* cabbages are large enough to start cutting. *Tundra* has a sweet, nutty flavour which makes it eminently suitable for eating raw, so rather than 'doctor' it by turning it into coleslaw, I decided just to shred it and toss it with the lettuce. Either John did not notice, or he agreed with me on the flavour, as nothing was said and he happily consumed the salad.

Thursday 4

The last few nights have been very frosty and where the sun does not reach, the frost has remained all day, so I thought it wise to lift some leeks and parsnips before the ground became too hard to get them out. This was the first digging of the *Winora* leeks, and they were surprisingly good, considering that the more than usually lush foliage of the parsnips has been overhanging them for several months. Most of them are at least 1in (2.5cm) thick, with about 9in (22.5cm) of white stem, even though they were earthed-up just the once. I am well pleased with these. In the world of vegetables, biggest is not always best.

At this time of year, leeks and parsnips will keep for quite a while after lifting. I have found the best way is to dig them up carefully with a flat-tined potato fork, damaging as little of the root as possible, and with plenty of earth still sticking to them. I then keep them in big plastic buckets in the garage, which is cool and light.

Even though I cook leeks with about 2in (5cm) of the green part still attached, I still feel it is a waste to discard the rest of the leaves. Apart from the very top part, which is generally tough and tatty, I save the rest and use it, chopped, within a short period, in soups, casseroles and savoury pies. As far as possible, therefore, I plan my menus so that most parts of the leeks can be used in one way or another.

I was not at all keen on leeks until I was well into adulthood and catering for myself. Possibly this is because when I was a child, my mother insisted on serving them, boiled up in plenty of water, without any accompaniments, apart from a liberal helping of grit. It is absolutely essential that leeks are washed until they are completely free from soil or they are most unpleasant to eat. Also, when earthing-up and working around the plants, care must be taken to make sure no soil falls onto the foliage, or it will gradually get down into the centre of the plant and it is a difficult job to wash it all out from the tightly packed leaves.

I cook leeks with as little water as possible – just enough to prevent them drying out completely. My favourite way is to braise them in the oven, dotted all over with butter and seasoned with salt and plenty of black pepper. The liquid which comes out of them can be used as the basis of a vegetable stock. Steamed leeks have a very good texture and flavour, and are excellent served with a cheese sauce.

Wednesday 10

I can leave it no longer – something must be done with that oversized *Greyzini* courgette. As it weighs in the region of 2lb (1kilo), it makes a good candidate for stuffing.

About once a week I try to have a meat-free day, supplementing our copious supply of vegetables with cheeses, eggs, fish and mushrooms. This year I have been growing mushrooms in a kit in the conservatory, and they have been much more successful than on previous occasions. If I leave them until they have opened out flat, I can usually manage to pick in the region of 1lb ($\frac{1}{2}$ kilo) at a time. As there were a nice few ready for gathering, mushroom-stuffed marrow seemed a good idea.

Mushroom-stuffed marrow

Ingredients
1 medium marrow

For the stuffing:
2 large shallots (or the equivalent in smaller ones), peeled and finely chopped

8oz (225g) mushrooms, wiped and finely chopped

1 tomato, skinned and chopped

2 tablespoons fresh breadcrumbs, preferably wholemeal

1 teaspoon dried sage or 1 tablespoon finely chopped fresh sage

1oz (25g) butter or margarine

Salt and freshly ground black pepper

Method
Melt the butter in a frying pan, add the shallots and fry for a minute or two until soft. Add the mushrooms and fry for a further two minutes. Combine the tomato, breadcrumbs, sage, salt and pepper and add to the shallot and mushroom mixture to make a stuffing, lightly pressing together with the fingers. If the mixture is too dry to hold together, stir in a small, lightly beaten egg.

Remove the skin from the marrow, cut off one end, and scoop out the seeds (in the case of our *Greyzini* marrow, there were not many seeds, so enough of the centre had to be removed to accommodate the stuffing). Fill the marrow with the mushroom stuffing and replace the cut-off end. Wrap in greased aluminium foil, place in a dish just large enough to accommodate the marrow and cook in a moderate oven at 160°C (325°F, gas mark 3) for $1\frac{3}{4}$ −2 hours.

Saturday 13

So far this winter I have not bothered with the Enviromesh netting, but today I was horrified to find that the pheasants had discovered my tender, young *Pilgrim* cauliflower plants and had turned the leaves into a collection of lacy mid-ribs. Winter cauliflowers are surprisingly tough, however, and they should recover in the spring if the damage gets no worse. I pulled out the few weeds growing in the row, broke up the soil a little between the plants, which has become very 'panned' with all the wet weather we have had during the past few weeks, and covered them with Enviromesh, supported with the plastic hoops which Agralan supply for converting Enviromesh into a cloche system. This will prevent the weight of the wet mesh flattening and damaging the plants where it touches. As the pheasants are obviously now on the lookout for a tasty snack, I thought it prudent to cover the Brussels sprouts and purple sprouting broccoli at the same time. The Enviromesh was wide enough to reach the row of rhubarb chard, so I covered this as well. This treatment should encourage the plants to produce a final crop of young leaves in the new year before I need to remove them and prepare the ground for the next season.

Sunday 14

Some of the younger members of John's family came for the day – John's youngest son, who refuses to eat vegetables but can sometimes be persuaded to toy with a small portion of carrots or roast parsnips; his wife, who enjoys anything and everything; and four children under eleven, who like many youngsters, tolerate vegetables to a greater or lesser degree depending upon type. Brussels sprouts are never a great children's favourite, but I find if you use the very small sprouts growing towards the top of the plant, (the 'mini-sprouts') and mix them with sliced baby carrots, such as the *Early Nantes* I froze during the summer, they become more acceptable. Young Josh, for some reason best known to himself, dislikes all roast meats, including chicken and turkey, so it was no traditional Sunday lunch for us this weekend, but a meat pie, which was apparently quite satisfactory. A trip to the newly opened local 'fun factory' was not exactly an adult's idea of unbounded pleasure, but the children enjoyed it, and it does happen to be right next door to my favourite garden centre, so it was an excuse to sneak off and pick up just one or two little things for the garden round the cottage. This is to be opened for two 'bulb Sundays' in the spring in aid of the local hospital MRI scanner appeal so it will need to be exceptionally interesting in March and April next year – even more so than usual, and it usually is quite pretty at that time of year.

Saturday 20

We are spending Christmas in Sheffield with John's brother and sister-in-law, which means I will not have a chance to cook the Christmas meal at home, so I thought John and I might have a leisurely pre-Christmas lunch today. We have spent a hectic week decking the halls with boughs of holly (well, putting up a few token decorations and adorning our new, artificial Christmas tree), posting hundreds of tardy cards and gasping with horror at all the people we have forgotten this year. In short, we have been finishing everything that needs to be finished before the whole country grinds to a halt for the best part of three weeks.

As I am sure we will have ample opportunity during the festive season to indulge in turkey, pork, chicken and ham in all their many guises, I thought we would treat ourselves to a duck. The 'new' *Rocket* potatoes I froze back in July, steamed with a sprig of mint from the pot I keep in the greenhouse for winter use, could almost have been freshly dug. We long ago used up all our frozen peas, but *Sugar Snap* are just as good an accompaniment to roast duck. As a second vegetable, I served the first of the *Tarvoy* savoys. Even John admitted that, as cabbages go, it was very pleasing. Unusually, I decided not to steam it, but cooked it for a few minutes in our new microwave oven, in a small amount of a half-and-half mixture of water and white wine vinegar, plus a generous knob of butter. When microwaving, the salt and pepper should be added to taste after cooking, otherwise the seasoning can become too strong. I always cook some of the dark green, outside leaves of the savoy, as they add flavour and are the most nutritious part of the vegetable.

Tuesday 23

When we spend Christmas at the in-laws', it is my job to supply the bulk of the vegetables. As we never know until we arrive how many people will be sitting round the table, or, indeed, just how many meals we shall participate in before we return home, the day before we leave is always spent in harvesting and cleaning what we hope will be an adequate supply. This year I am taking a *Tundra* cabbage, *Tarvoy* savoy, two sticks of sprouts and plenty of leeks and parsnips. They will stay fresh longer if most of the preparation is done when they are needed, but they were in such a dirty mess that I decided to trim the outer leaves off the cabbages, remove the slimy leaves from the parsnips and trim the tops of the leeks before giving them a good hosing down prior to packing them loosely in dustbin bags. Polythene bags are not suitable for long-term storage of vegetables as they soon sweat and start to rot, but it is more convenient to transport them in something waterproof. Once we arrive, we can take them out and pack them in something more appropriate. Our silt soil is a wonderful growing medium, but in winter, after frost and rain, it is quite the most claggy I have ever encountered, and cold, too, when it sticks to the fingers.

Thursday 25

Christmas Day was just as good as usual, lots of presents, plenty to eat and drink, and a chance to meet up with friends and relatives we have not seen in a long time. At this time of year, cookery magazines are overflowing with different, often elaborate and fiddly, ideas for serving turkey and seasonal vegetables, but to me, Christmas is a time for tradition, and it is hard to improve on a well-roasted, traditional bird with traditional accompaniments and all the traditional kinds of traditionally-cooked, fresh winter vegetables – a feat which in itself is not as easy as it sounds! Ours today was just such a meal, enjoyed by all, but brother-in-law, Fred, as usual, just had to be different. He does not like fresh vegetables at all, apart from potatoes that is, so his Christmas lunch was – roast turkey and baked beans! Ah, well, it takes all sorts ...

Wednesday 31

The end of the year is always a period for looking back. Amongst other things, this New Year's Eve is a good time to reflect on the successes – or otherwise – of the garden over the last twelve months, and, in particular, the vegetable garden which now occupies a sizeable amount of my time.

There is no doubt that growing your own vegetables is worthwhile – the food is fresher, cheaper, and you know exactly what has been used in the way of chemicals. In general, the flavour is better and you eat far more than if you were buying in, so you have no worries about those five beneficial portions of fruit and vegetables nutritionists are now telling us are a must for healthy living. Your vegetable consumption will far exceed that amount once the kitchen garden is established! The physical work involved in producing good crops ensures you get fresh air and plenty of exercise – so much more tangibly productive than jogging or 'working out'!

Of course, home vegetable production has its drawbacks. Some crops do fail and have to be resown or replanted – but this can happen to commercial growers, too. It is not realistic to expect every single item to look as perfect as those you encounter on the greengrocer's shelves, but a lot will be comparable, and, well cooked and served, the end result will be as good, if not better. Much thought, planning and energy must be expended on a successful kitchen garden, and there are times when you would rather do anything than dig, plant, hoe and gather. Faced with a pile of dusty lettuce, muddy leeks or soil-encrusted potatoes, it is understandable that your thoughts often drift wistfully to the rows of immaculate produce on the supermarket shelf, but remember the nutritional value will always be higher in freshly harvested crops. You have total control over the choice of every variety, and good gardening practice will help you to make the most of the aids to cultivation that Nature provides. But if you were looking for one reason above all others to grow your own produce, it is that you will achieve something you will never buy – the total satisfaction of producing your own wholesome, safe, high-quality provender, secure from food scares and governmental whims. And if that isn't a good argument for doing it yourself, I don't know what is!

Looking Forward

Friday 1 January

So, here we are at the start of another year. In the kitchen garden, we still have plants of *Tundra* and *Tarvoy* cabbage; enough *Icarus* Brussels sprouts to last us until February; maturing plants of early purple sprouting broccoli *Rudolph*, which will be ready for picking shortly; parsnips and leeks to see us through to the spring and, for late spring and early summer, *Walcheren Winter Pilgrim*, the winter-hardy Japanese onions.

The *Jet Set* onions are almost used up, but there are plenty of shallots left which, as far as I am concerned, will do the same job. If they begin to show signs of sprouting, I shall freeze the rest so I will have enough to last me until the Japanese onions are ready – not forgetting, of course, to save enough to plant in March for the coming season's crop. These will be started into growth in trays during February.

In the freezer, there are French and runner beans, cauliflower and broccoli, carrots, turnips, mangetout and snap peas, broad beans, tomatoes and parsley, enough to last, with care, until the new season. In March I will need the area at present occupied by the leeks and parsnips for salads, onions and tomatoes. If there are any left, they can be dug up and 'heeled in' in a spare piece of ground between the vegetable garden and the orchard for a month or so to keep them fresh. At the end of this time, if they have still not been used, I shall freeze them; they will be very useful for soups, stews and casseroles in the summer when they are out of season. It will be possible to leave the purple sprouting broccoli *in situ* until it has been harvested, then the site will be dug and prepared for root vegetables which, if necessary, can be sown slightly later in April to accommodate the broccoli.

As parts of the kitchen garden continue to become vacant, they can be dug over and, later in the winter, prepared and covered over for the second year of our crop rotation. I have plenty of compost which can be trenched in at the time, but a balanced fertilizer will not be required until a week or two before sowing time.

It is pleasant to know that the seeds for next year are already in the kitchen cupboard at the Patch. Next week, once the Christmas decorations are dismantled, the cards packed away in boxes ready for whichever charity asks for them first, and life has otherwise returned to normal for another eleven months, I shall obtain my seed potatoes and once more put them in trays in the shed to sprout.

Possibly the best thing about vegetable growing is that, once started, there is no beginning and no end, just preparation, planting and harvesting in due season. However many times we sow, and on however many occasions we harvest, there is always the same thrill on spotting the first seedlings to break through the earth, and the same delight at the first taste of each new crop of the year. And long may it remain that way.

APPENDIX 1

Blanching and cooling times for freezing vegetables

Blanching and cooling times are equal.

Peas, mangetout and beans may be frozen for up to eight weeks without blanching without deterioration in texture and flavour.

Onions, Brussels sprouts, cauliflower, broccoli and cabbage may be frozen for one month without blanching. Do not exceed this period as serious deterioration in flavour will occur.

Vegetable	Time (minutes)
Beans, broad	$1\frac{1}{2}$
Beans, dwarf or climbing (whole) (cut)	3 2
Beetroot	Boil until tender, freeze when cold
Broccoli	3
Brussels sprouts	3
Cabbage (shredded)	$1\frac{1}{2}$
Carrots (whole, young) (sliced, for soups and casseroles)	3 2
Cauliflower (florets)	3
Courgettes (small, whole or larger, sliced)	$1\frac{1}{2}$
Leeks (cut into 1in [2.5cm] pieces)	2
Onions (sliced)	2
Parsley	Freeze dry without blanching and crumble while frozen

Parsnips (diced, for soups, stews and mash)	
Peas, shelled	$1\frac{1}{2}$
Peas, mangetout and sugar snap	2
Potatoes (new, small)	4 NB: skins will rub off easily after blanching
Spinach, chard, spinach beet	$1\frac{1}{2}$. Alternatively, steam until cooked, press out excess moisture, cool, freeze and reheat when required.
Sweetcorn	3–6 minutes, according to size
Tomatoes	Freeze without blanching; use for soups, stews and other recipes requiring cooked tomatoes. Alternatively, simmer in own juice, pass through a sieve and freeze in small tubs. Use in recipes calling for tomato purée.
Turnips (diced)	2

Winora *leeks usually have a good length of white stem*

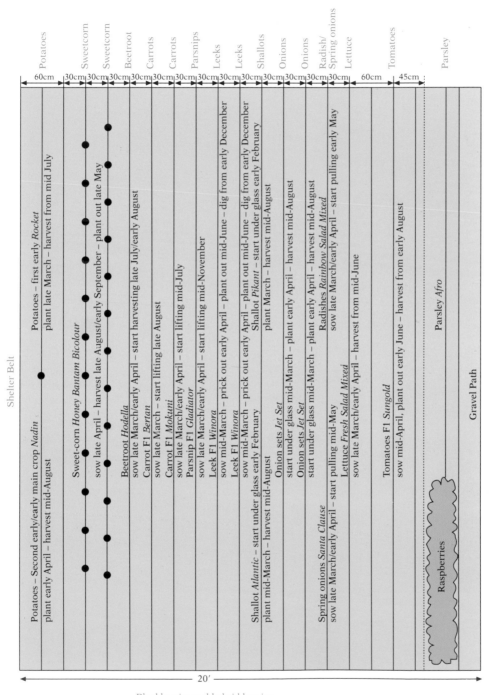

Shelter Belt

Potatoes	Sweetcorn	Sweetcorn	Beetroot	Carrots	Carrots	Parsnips	Leeks	Leeks	Shallots	Onions	Onions	Radish/Spring onions	Lettuce	Tomatoes	Parsley
60cm	30cm	30cm	30cm	30cm	30cm	30cm	30cm	30cm	30cm	30cm	30cm	30cm	30cm	60cm	45cm

Potatoes – first early *Rocket*
plant late March – harvest from mid July

Potatoes – Second early/early main crop *Nadin*
plant early April – harvest mid-August

Sweet-corn *Honey Bantam Bicolour*

sow late April – harvest late August/early September – plant out late May

Beetroot *Hodella*
sow late March/early April – start harvesting late July/early August

Carrot F1 *Bertan*
sow late March – start lifting late August

Carrot F1 *Mokuni*
sow late March/early April – start lifting mid-July

Parsnip F1 *Gladiator*
sow late March/early April – start lifting mid-November

Leek F1 *Winora*
sow mid-March – prick out early April – plant out mid-June – dig from early December

Leek F1 *Winora*
sow mid-March – prick out early April – plant out mid-June – dig from early December

Shallot *Atlantic* – start under glass early February
Shallot *Pikant* – start under glass early February
plant mid-March – harvest mid-August
plant March – harvest mid-August

Onion sets *Jet Set*
start under glass mid-March – plant early April – harvest mid-August
Onion sets *Jet Set*
start under glass mid-March – plant early April – harvest mid-August

Radishes *Rainbow Salad Mixed*
sow late March/early April – start pulling early May

Spring onions *Santa Clause*
sow late March/early April – start pulling mid-May
Lettuce *Fresh Salad Mixed*
sow late March/early April – harvest from mid-June

Tomatoes F1 *Sungold*
sow mid-April, plant out early June – harvest from early August

Parsley *Afro*

Gravel Path

Raspberries

20'

Blackberries and hybrid berries

Courgettes Salad Collection – sow early April – plant out late May

156

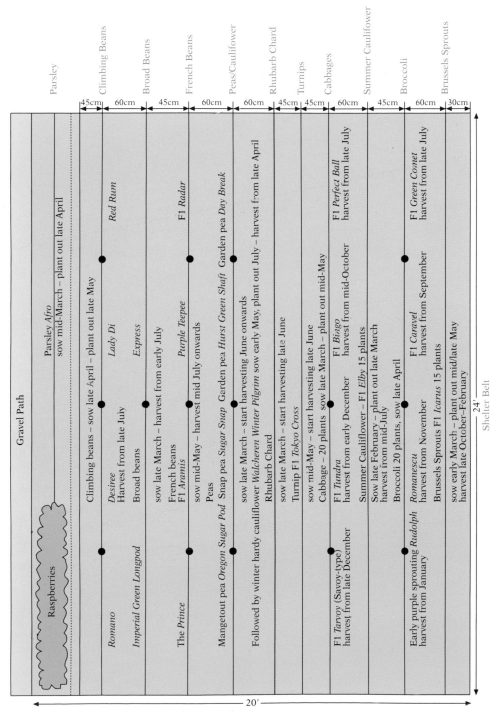

Shelter Belt

Blackberries and hybrid berries

A VEGETABLE GARDEN CROP ROTATION

First Year

Brassicas	Others		Others	Roots
Brussels sprouts	Rhubarb Chard		Tomatoes	Parsnips
Broccoli	Peas		Lettuce	Beetroot
Cauliflowers	French Beans		Spring onions	Sweet corn
Cabbage	Runner/climbing		Radishes	Potatoes
Turnips	beans		Onions	
			Shallots	
			Leeks	

N (arrow pointing down/southeast)

Parsley | Path | Parsley

Second Year

Roots	Brassicas		Others
Parsnips	Brussels sprouts		Tomatoes
Carrots	Broccoli		Lettuce
Beetroot	Cauliflowers		Spring Onions
Sweetcorn	Cabbage		Radishes
Potatoes	Turnips		Onions
			Shallots
			Leeks

Parsley | Path | Parsley

Third Year

Others		Roots	Brassicas
Tomatoes		Parsnips	Brussels sprouts
Lettuce		Carrots	Broccoli
Spring onions		Beetroot	Cauliflowers
Radishes		Sweetcorn	Cabbage
Onions		Potatoes	Turnips
Shallots			
Leeks			

Parsley | Path | Parsley

APPENDIX 2

Vegetable varieties for second and third years

SECOND YEAR

Broad beans

Aquadulce Claudia   A popular old favourite which can be sown very early and therefore produces a crop well in advance of most others.
Verdy   Superb flavour and a very high yield makes this new variety well worth trying.

Dwarf beans

Atlanta   A flat-podded, stringless variety with a rich flavour, which can be sliced or served whole.
Capitole   Produces hundreds of short, round, slim pods, which are eaten whole and freeze well. Resists virus and tolerates high temperatures.

Runner/climbing beans

Kentucky Blue   A very early climbing French bean with long, round pods and a long cropping season.
White Lady   A new, white-flowered bean with a delicious flavour.
Scarlet Emperor   A popular runner bean with huge crops of very long, tasty beans.

Beetroot

F1 Red Ace   A strong grower with uniform, well-shaped bulbs, capable of coping with dry weather conditions.

Brassicas

Broccoli

F1 Trixie   A stocky, vigorous variety with large, firm, domed heads. Club root resistant.

Temple   Late-maturing '*Romanesco*' type with pointed heads on medium-sized heads. Suitable for close planting and therefore useful for small gardens.

Sprouting broccoli *White Eye*   An early-heading variety with attractive-looking white spears.

Brussels sprouts

F1 Cavalier   Can be grown as an early variety or sown later for a crop in late winter. Firm, uniform, well-coloured sprouts which freeze well.

Cabbage

Summer *F1 Hispi*   An award-winning, hybrid F1 summer cabbage which has been around for many years and has stood the test of time. Excellent for close planting and small gardens, slow to run to seed.

Autumn *F1 Minicole*   Another well-established hybrid variety for the small garden. Small, uniform heads will stand for months without running to seed.

Winter *F1 Red Savoy*   A superb combination of size, appearance and colour. The red veins turn blue on cooking, the flavour is sweet, and the compact size is just right for the small family.

Cauliflower

Summer *F1 Violet Queen*   The curd turns green when cooked. The flavour is first-rate, and the purple florets are ideal for novelty salads and crudities.
Winter *F1 Plena*   An adaptable variety with good-quality, dense curds which freeze well.

Carrot

F1 Ingot   For an early crop. High yielding, long roots which are well flavoured, sweet with a high vitamin A content, so excellent for eating raw. Freezes well.
F1 Liberno   Main crop – harvest summer and autumn. Good flavour, both raw and cooked, and resistant to splitting, so a good choice if dry periods are likely.

Chard (leaf beet)

Lucullus   A prolific cropper, with dark green leaves and wide, white mid-ribs.

Courgette

F1 Sardane   High yielding, early to mature, with well-shaped, dark fruit which are easy to pick because they are produced on long stalks.
F1 Supremo   Early maturing, highly productive variety with a strong resistance to virus which is now often a problem after our increasingly mild winters.

Leek

F1 Upton   Produces strong, uniform plants with a high resistance to leek rust.

Lettuce

Little Gem   A miniature cos with high-quality, sweet-flavoured heads. This is possibly still my favourite lettuce of all time. Supermarkets apparently agree as it is now widely available, although at an exhorbitant price.
Red Lettuce Mixture   A range of red-leaved varieties maturing over a long period, useful for visual effect in salads.

Onions and shallots – **from sets**

Golden Ball   A variety which prefers to be planted late. This helps to space out planting times at the busiest period of the year in the vegetable garden.
Atlantic and *Pikant*   I honestly do not think I can improve on the shallots I grew last season. The golden and red varieties both produced an enormous crop, and the variation in colour proved to be greatly beneficial in recipes.

Spring onions

Winter White Bunching   Produces masses of onions and can be sown in spring for a summer crop, or in late summer for autumn and winter.

Parsnip

Tender and True   A long, good-looking variety with a fine flavour, popular with vegetable growers for many years.

Peas

Kelvedon Wonder   An old favourite, early, good cropping, well flavoured and tolerant of disease.
Rondo   A double-podded, maincrop variety with an average of 10 peas per pod, which is fusarium wilt resistant and of a very high flavour.
Norli   A mangetout variety, which, unlike many, does not grow tall and therefore needs little support.

Potatoes

First early
Maris Bard   Very early, this is a popular choice with farmers because of its heavy crops, good flavour and resistance to blight.

Second early
Estima   A heavy-cropping, modern variety with yellow, waxy flesh and a well-shaped tuber, which has proved to be easy and reliable and is therefore gaining in popularity with commercial growers. It stores well and can be left in the ground to harvest as a maincrop variety.

Radishes

Prinz Rotin   A variety I have grown, on and off, for many years and, in my opinion, one which beats many newer ones hands down.

Stands well, does not split in erratic rainfall conditions and can be allowed to grow large without losing flavour and texture.

Sweetcorn
F1 Champ   A super-sweet, early maturing variety suitable for poor summers and northern climates.

Tomato
Brigade   A wilt-resistant tomato producing medium-sized tempting fruit suitable for sun-drying, eating fresh, making into purée or freezing.

Turnip
Snowball   A popular, early turnip with a sweet flavour and good texture.

Path edging

Basil
Bush   A popular flavouring for fish, salads and tomato dishes.
Lemon   An unusual flavouring for fish and certain meats.
Sweet green   The flavour is reminiscent of cloves and mint. Used in pasta dishes and to flavour courgettes. Can be used fresh or dry in salads, casseroles and flavoured vinegars.
Cinnamon   An ornamental form which has cinnamon-coloured stems, purple flowers and dark green leaves, with a hint of cinnamon flavouring.

Parsley
Clivy   Dwarf and neat.
Envy   Tightly curled and deep green.
Par-cel   Looks like parsley and tastes like celery.

THIRD YEAR

Broad beans

The Sutton   An early-cropping bean which gives a good yield of small pods on short, stocky plants. Suitable for small gardens, exposed situations and close cropping.

Dwarf beans

Vilbel   A newish bean I tried when just launched and found really good. It is a 'filet' type, producing masses of fine, round pods throughout the summer.

Masai   Very early to crop, this is another pencil-podded variety with huge yields of splendid flavour.

Runner/climbing beans

Streamline   Vigorous growth, early to crop, with long pods produced in bunches over a very long season.

Royal Standard   Fleshy, stringless pods up to 20in (50cm) long with a good setting capacity even in difficult weather conditions.

Red Rum   A variety grown in the first year, which was possibly the best of a good collection, both for flavour and yield, and is therefore well worth growing again.

Beetroot

Detroit 6 – Rubidus Globe   A variety which is well-shaped, heavy cropping, and very slow to 'bolt'.

Brassicas

Broccoli

F1 Sprouting Claret   Vigorous growth and plenty of red spears in late spring and early summer, an unusual time for fresh sprouting broccoli.

F1 Caravel   This was grown in year one, and proved to be popular with everyone. Large, domed heads and plenty of side shoots in summer and autumn. Plants can be grown close together for a high yield from a small area.

Romanesco   Also grown in year one, this is still the classic early winter broccoli with the characteristic pointed spears.

Brussels sprouts

F1 Trafalgar A late maturing variety producing a heavy, uniform crop of medium-sized sprouts with a sweet flavour.

Cabbage

Summer *F1 Quickstep* Very early, with a solid head and few outer leaves.
Autumn *F1 Bingo* Grown in year one, this was a fine cabbage for both coleslaw and pickling.
Winter *F1 Marabel* A hardy *January King* hybrid which will stand well until the end of March.

Cauliflower

Winter *Walcheren Winter 3 – Armado* Winter hardy, producing heavy crops of large, white close heads in late spring.
Summer *Dok* A classic variety it is hard to beat. The leaves curl naturally over the high-quality curds for protection.

Carrot

Suko A 'finger carrot' which is very fast-maturing and is harvested small. Sweet flavour and ideal for freezing whole.
Lange Rote Stumpfe 2 Zino The complete opposite of Suko. Produces enormous, juicy maincrop roots which can be overwintered in the ground, in sand or in clamps.

Chard/leaf beet

Perpetual spinach (spinach beet) A good alternative to Swiss chard as it does not easily run to seed the first year and is hardy, so a final crop can be harvested the following spring when leaf vegetables are scarce.

Courgette

F1 Defender An exceptionally heavy cropper with mid-green fruits. Resistant to cucumber mosaic virus.
F1 Gold Rush One of year one's courgettes, the lovely colour of which is so striking when sliced and served raw in salads.

Leek

Toledo Very long stems; a high yielding crop which lasts through winter and spring into May (if they have not all been eaten long before then!).

Lettuce

Blush   A mini-iceberg lettuce suitable for the small family and for close cropping – each lettuce serves one person. Deep green outer leaves tinged red.

Sherwood   A compact, semi-cos type with a flavour said to be superior to *Little Gem* (I think they are equal!). Pelleted seed makes for ease of sowing; the total crop should be around 30 plants.

Onions and shallots

For a change, I shall be growing them from seed this year.

Onions

My friend Jenny paid an enormous amount of money in our local supermarket for three pre-packed onions of different colours recently. I shall be able to supply her with all she needs with the following varieties.

F1 Albion   A tasty onion with a globular, uniform shape and snow-white skin.

F1 Caribo   Heavy cropping with good keeping qualities; the skins are deep straw-gold.

Red Baron   Attractive, dark red-skinned semi-globe onion with red-tinged flesh.

Spring onions

White Lisbon   This was the variety I filled in with in year one after patchy germination of *Santa Clause*. It was mild, reliable and long standing, and is possibly still the best salad onion for general use.

Parsnip

Avonresister   A canker-resistant variety which requires less thinning to produce a crop of uniform, well-shaped roots.

Peas

Little Marvel   An old variety which is still popular. Early, sweet and compact.
Hurst Green Shaft   I am sure I shall be glad to get back to my favourite pea after a year's break.
Sugar Gem   Stringless and powdery mildew resistant.

Potatoes

First early

Pentland Javelin   A well-established early potato which is not as quick to mature as some, but is an exceptionally heavy cropper with oval, white-fleshed tubers, a waxy texture and a resistance to scab and eelworm.

Second early

Wilja   A newer but proven variety which, with *Nadine* and *Estima*, is probably the most dependable of the second earlies to date. High yielding and cooks well.

Radishes

F1 Juliette   Uniform, large, quick maturing with a mild flavour.

Swede

Best of All   I think I might fancy a change from turnips in the third year as swedes are such a useful winter root vegetable. This variety is hardy, dependable and heavy yielding, with a purple skin and yellow flesh.

Sweetcorn

F1 Butterscotch   A super-sweet variety with long cobs and a good cold tolerance.

Tomato

Gardeners' Delight   An old variety of cherry tomato with an old-fashioned flavour and a heavy crop of small, brightly coloured fruits.

Path edging

On the assumption that the freezer will still contain enough parsley for cooking for the next twelve months, and parsley for fresh garnishes can be provided by growing one or two plants in pots at home, it might be nice to have a total change of subjects for the edging to the path dividing the two halves of the kitchen garden.

Flower Petal Salad   A mixture of different flowers with edible petals which not only make a decorative edging to a kitchen garden path, but will add interest to salads and can even be used as a garnish for summer soups.

Rocket

An aromatic leaf vegetable which is picked regularly while young and tender and used in mixed green salads, chopped in dips and dressings, or cooked like spinach.

APPENDIX 3

Approximate yields of crops grown in 24ft (7m) rows

Rhubarb chard **25lb** (11$\frac{1}{8}$ kilo)

Broad beans
Express, Imperial Green Longpod
Total shelled weight of both varieties 12lb (5$\frac{1}{2}$ kilo)

Climbing beans
Red Rum, Lady Di, Romano, Desiree
Total weight of all varieties 100lb (45$\frac{1}{2}$ kilo)

Dwarf beans
Aramis, Purple Teepee, Cropper Teepee, The Prince, Radar
Total weight of all varieties 12lb (5$\frac{1}{2}$ kilo)

Radar *dwarf beans*

Beetroot
Mondella 10lb ($4\frac{1}{2}$ kilo)

Broccoli
F1 Green Comet 5 plants ($\frac{1}{4}$ row) 7lb ($3\frac{1}{4}$ kilo)
F1 Caravel 5 plants ($\frac{1}{4}$ row) 7lb ($3\frac{1}{4}$ kilo)
Romanesco 5 plants ($\frac{1}{4}$ row) 8lb ($3\frac{1}{2}$ kilo)

Early Purple Sprouting
Rudolph ($\frac{1}{4}$ row) 9lb (4 kilo)

Brussels sprouts
F1 Icarus 15 plants 38lb ($17\frac{1}{4}$ kilo) plus 5lb ($2\frac{1}{4}$ kilo) sprout tops (used like cabbage)

Cabbage
F1 Perfect Ball 5 plants 14lb ($6\frac{1}{8}$ kilo)
F1 Bingo 5 plants 15lb ($6\frac{3}{4}$ kilo)
F1 Tundra 5 plants 18lb (8 kilo)
F1 Tarvoy 5 plants 14lb ($6\frac{1}{8}$ kilo)

Cauliflower
F1 Elby 15 plants 25lb ($11\frac{1}{8}$ kilo)
Walcheren Winter Pilgrim 15 plants 30lb ($13\frac{1}{2}$ kilo)

Carrots
F1 Mokum, Early Nantes 8lb ($3\frac{1}{2}$ kilo)
F1 Bertan 15lb ($6\frac{3}{4}$ kilo)

Courgettes
De Nice à Fruit Rond, F1 Gold Rush, F1 Greyzini, F1 Sardane
Total from 20 plants 60lb ($27\frac{1}{4}$ kilo)

Leek
Winora 18lb (8 kilo)

Lettuce
Fresh Salad Mixed, Little Gem
About 25 mature heads from three successional sowings, plus thinnings

Onion sets
Jet Set 45lb ($20\frac{1}{2}$ kilo)

Shallots
Atlantic, Pikant
Total for both varieties 27lb (12¼ kilo)

Salad onions
Santa Clause, White Lisbon
10 ft (3m) row. About 170 onions from two sowings.

Jet Set *onions*

Parsnips
F1 Gladiator 40lb (18 kilo)

Peas
Daybreak 6ft (1.8m) row. Shelled weight 2lb (just under 1 kilo)
Hurst Green Shaft 6ft (1.8m) row. Shelled weight 4lb ($1\frac{3}{4}$ kilo)
Oregon Sugar Pod 6ft (1.8m) row. 8lb ($3\frac{1}{2}$ kilo)
Sugar Snap 6ft (1.8m) row. 15lb ($6\frac{3}{4}$ kilo)

First early potato
Rocket 12lb ($5\frac{1}{2}$ kilo)
Second early/early maincrop potato
Nadine 18lb (8 kilo)

Radish
10ft row. About 90 roots from two sowings.

Sweetcorn
F1 Honey Bantam Bicolour 27 cobs from 21 plants

Tomato
F1 Sungold 68lb (31 kilo)

Turnip
F1 Tokyo Cross 16lbs ($7\frac{1}{4}$ kilo) plus turnip tops from young thinnings, cooked like spinach

INDEX

Page numbers in italics indicate illustrations

P146 Environmesh
Agralan

Maxiwrap P47 Seaweed based
fertiliser

Metpost Garden Stakes P52